In Search of Dark Matter

Ken Freeman and Geoff McNamara

In Search of
Dark Matter

 Springer

Published in association with
Praxis Publishing
Chichester, UK

Professor Ken Freeman
Research School of Astronomy & Astrophysics
The Australian National University
Mount Stromlo Observatory
ACT
Australia

Mr Geoff McNamara
Science Teacher
Evatt
ACT
Australia

SPRINGER–PRAXIS BOOKS IN POPULAR ASTRONOMY
SUBJECT *ADVISORY EDITOR*: John Mason B.Sc., M.Sc., Ph.D.

ISBN 10: 0-387-27616-5 Springer Berlin Heidelberg New York
ISBN 13: 978-0-387-27616-8

Springer is a part of Springer Science + Business Media (*springeronline.com*)

Library of Congress Control Number: 2005931115

Cover design: Jim Wilkie
Copy editing and graphics processing: R. A. Marriott
Typesetting: BookEns Ltd, Royston, Herts., UK

Printed in Germany on acid-free paper

Table of contents

Authors' preface

Although science teachers often tell their students that the periodic table of the elements shows what the Universe is made of, this is not true. We now know that most of the Universe – about 96% of it – is made of dark material that defies brief description, and certainly is not represented by Mendeleev's periodic table. This unseen 'dark matter' is the subject of this book. While it is true that the nature of this dark matter is largely irrelevant in day-to-day living, it really should be included in the main-stream science curricula. Science is supposed to be about truth and the nature of the Universe, and yet we still teach our children that the Universe is made up of a hundred or so elements and nothing more.

Dark matter provides a further reminder that we humans are not essential to the Universe. Ever since Copernicus and others suggested that the Earth was not the centre of the Universe, humans have been on a slide away from cosmic significance. At first we were not at the centre of the Solar System, and then the Sun became just another star in the Milky Way, not even in the centre of our host Galaxy. By this stage the Earth and its inhabitants had vanished like a speck of dust in a storm. This was a shock. In the 1930s Edwin Hubble showed that the Milky Way, vast as it is, is a mere 'island Universe' far removed from anywhere special; and even our home galaxy was suddenly insignificant in a sea of galaxies, then clusters of galaxies. Now astronomers have revealed that we are not even made of the same stuff as most of the Universe. While our planet – our bodies, even – are tangible and visible, most of the matter in the Universe is not. Our Universe is made of darkness. How do we respond to that?

The last fifty years have seen an extraordinary change in how we view the Universe. The discoveries that perpetuated the Copernican revolution into the twentieth century have led to ever more fundamental discoveries about how the Universe is put together. But parallel to the discovery of the nature of our Galaxy and galaxies in general ran a story almost as hidden as its subject. The laws of gravity that Newton and later Einstein propounded were put to good use in discovering new worlds in our Solar System, namely Neptune and Pluto. These same techniques – of looking for the gravitational effects on visible objects by unseen objects – led astronomers to realise that there exists much more matter than we can see. This book tells that story. It is a story of false trails that ultimately pointed in the right direction; of scientists' arrogance and humility, curiosity and puzzlement. But most of all it is a story that shows the persistent nature of science and scientists who consistently reveal just how much more there is to learn.

The problem is that each new discovery seems to show not more about the Universe, but simply how much we have yet to learn. It is like a person who wakes in a dark cave with only a candle to push back the darkness. The feeble glow reveals little but the floor of the cave and the surrounding darkness. Hope rises when a torch is found; but the additional luminance does not reveal the walls of the cave, rather the extent of the darkness. Just how far does the darkness extend? We have yet to find out. This book describes how far into the night we can currently see.

This is also a story about science and scientists. All but one of the contributors is a scientist with expertise in specific aspects of the dark matter problem. The non-scientist of the group is Geoff McNamara, a teacher and writer, who was responsible for bringing the story together from the various contributors. Most of the historical and contemporary astronomical research into the location and quantity of dark matter was related by Ken Freeman, who has a long career in dark matter research since its revival in the late 1960s. Professor Warrick Couch, Head of the School of Physics at the University of New South Wales, relates how gravitational lensing has evolved into a technique that is now used to help map out the location and amount of dark matter in galaxies and galaxy clusters. The story of the exotic particles that perhaps make up dark matter is told in Chapters 11, 12 and 13. These chapters rely heavily on technical input from Professor Ray Volkas of the School of Physics, University of Melbourne, and his advice on particle astrophysics is gratefully acknowledged. Finally, Dr Charley Lineweaver of the Research School of Astronomy & Astrophysics, Australian National University, relates the implications of dark matter and the relative newcomer – dark energy – for the long-term fate of the Universe.

How do students react when their insignificance in time, space and now matter is revealed to them? As the immensity of the Universe is revealed, as the unimaginable distances in time and space become apparent and they realise they are not even made of the same stuff as the rest of the Universe, they feel small and insignificant. But this phase soon passes, and curiosity takes over. Students with very different academic ability and understanding of things astronomical all come to the same point: they want to know more. These young people are all scientists at heart – even if only a few will have the opportunity to pursue the subject professionally. It is for our students and like-minded readers everywhere that this book has been written.

Kenneth Freeman, FAA, FRS, Duffield Professor, Australian National University
Geoff McNamara, Canberra, ACT

List of illustrations

Prologue

The quest for darkness

Astronomy was once a quest for light. For millions of years, humans stared wide-eyed at the night sky, trying to piece together the nature of the Universe in which we live. But because of the limitations of the naked eye, the vast majority of our ancestors never suspected, and none knew, that the stars were other suns and the planets other worlds. Such revelations had to wait until the invention of the telescope – an instrument that simultaneously created and fulfilled the possibility of seeing fainter and more distant objects in the Universe. While the earliest telescopes were capable of little more than today's toy telescopes, they nonetheless revealed for the first time the moons of Jupiter, craters on our own Moon, and the myriad of fainter stars in the Milky Way. As telescopic power grew it was assumed that telescopes, being light-gatherers, would reveal ever more of the Universe that surrounds us, and that they would eventually reveal everything. Indeed, modern telescopes have provided us with images of objects so distant that they are not only close to the edge of the Universe, but almost at the edge of physical detectability. The interpretation of what telescopes reveal aside, without their light-gathering capability our understanding of the Universe might never have progressed beyond the Milky Way.

Our story begins in the first few decades of the twentieth century, when the first truly modern telescopes were built atop high mountains. Sitting in their new, ivory-coloured towers, astronomers were literally and metaphorically closer to the stars than they had ever been before. These were heady times for astronomers: our parochial view of the Universe in which the Milky Way was the dominant feature expanded to one where our Galaxy played a minor role. Stars gathered into galaxies; galaxies into clusters of galaxies. The general expansion of the Universe was independently and simultaneously discovered and explained, and astronomy and physics forged a partnership that is now inseparable: astronomers turned to physics for explanations of what they saw, and physicists realised the Universe was a laboratory of immense size and energies.

Despite the philosophical ramifications of these discoveries, or perhaps because of them, the future of astronomy looked bright. To astronomers, the night was ablaze with the light of uncountable suns at almost immeasurable distances. But everywhere astronomers looked, something was missing. The stars and galaxies sparkled in the night as far as the eye could see, like moonlight off an ocean, but their behaviour was peculiar: rather than huddling together in the cold emptiness of space bound by their own gravity alone, the stars and galaxies seemed to be pulled this way and that by dark, unexplained currents that seemed to permeate the cosmos. Far from dominating the Universe, the stars and galaxies behaved as if they were mere flotsam on a cosmic sea.

The problem is as simple to understand as it has been difficult to solve. There is only one way to interpret gravity, and that is the existence of matter. Everything in the Universe has a gravitational pull on everything else – a phenomenon that holds solar systems, star clusters, and galaxies together. The dynamics of the stars and galaxies hinted at more matter than meets the eye. But when astronomers tried to find the source of the gravity they found... nothing. As larger telescopes penetrated deeper into space they revealed structures on increasingly larger scales, yet every turn of the telescope revealed more of the same unexplained gravitational influence. Not only that, but the larger the scale – from stars to galaxies to clusters of galaxies – the greater the mysterious effect seemed to be. The further astronomers looked, the less of the Universe they saw.

Because of its invisibility, the unseen matter was once called 'missing mass'; but this is not a good term, since the location is well known, and astronomers can literally point their telescopes to it. Yet to even the largest, most powerful telescopes it remains invisible against the blackness of space, and so it has become known as 'dark matter'. However, this term understates the significance of the concept it represents. The dark matter mystery has evolved from simply another unsolved astronomical problem to one of the most important cosmological questions of all time. One reason is that despite the fact that it seems to outweigh visible matter by as much as a hundred times, no-one knows what dark matter is made of. Is this simply a limitation of the way we observe the Universe? Perhaps. But keep in mind that when dark matter was originally detected astronomers were limited to using optical telescopes; that is, they saw the Universe only in visible light. In the intervening seventy years or so, the Universe has been studied in a myriad of new wavelengths, each revealing new forms of previously invisible matter, including interstellar gas and dust, neutron stars, radio galaxies and black holes. But the addition of these previously unseen sources of matter falls a long way short of accounting for the effects of dark matter.

It could turn out that all this time we have simply been looking for the wrong kind of matter. The very term 'dark matter' implies matter that is non-luminous, simply not giving off any light. What if it is not even made of baryonic matter – the familiar protons and neutrons that make up stars and planets? (See Appendix 1.) Perhaps we should be looking for non-baryonic matter – exotic particles, many of which have yet to be discovered. Perhaps most of the Universe is not made of the same stuff as we are. Such a revolution in our thinking would not be unprecedented. Four hundred years ago the so-called Copernican Revolution displaced the Earth from the centre of the Universe. As has been noted by David

Schramm, we could now be experiencing the ultimate Copernican Revolution, in that not only are we not at the centre of the Universe, we may not even be made of the same stuff as most of the Universe.

There are many candidates lining up for the role of dark matter: neutron stars, primordial black holes, dead stars, neutrinos, and a whole family of exotic particles called WIMPs. We shall take a look at each of these and other candidates in turn, and see how scientists are trying to find them. However, we need to be careful about what conclusions we draw about the nature of dark matter. Astronomers are very creative storytellers, and can always construct an hypothesis to fit the facts; and the fewer facts available, the easier it is to fit the hypothesis. As astronomers grope in the darkness towards a fuller understanding of an astronomical problem it is important to invoke a principle known as Occam's Razor: the simplest – and usually most elegant – explanation is the one that is to be preferred. In the case of dark matter, this means it is better to assume that there is one sort of dark matter to account for the gravitational effects seen at Galactic and extragalactic scales. However, as our story unfolds you will see that it is more likely that things are not as simple as that. In fact, we might have to learn to live with several different sorts of dark matter, each providing the gravitational influence we see on different scales. Whatever it is made of, dark matter certainly played a role in the origin of the Universe. Without it, the Universe would have no galaxies, no stars, and possibly no-one to wonder why. Yet it does have them, and here we are.

Just as dark matter played a crucial role in the origin of the Universe, it may be a major factor in the cosmological tug-of-war between the expansion of the Universe and its self-gravitation. The expansion of the Universe – the implication of which was the Big Bang, the primordial fireball which gave birth to the Universe – was revealed around the same time as the discovery of dark matter. This expansion is struggling against the gravitational pull of the matter it contains. If the Universe contains too little mass, it will expand forever; too much and it will one day collapse in on itself again. Between these two extremes is perfect balance between gravity and expansion – a 'critical density' that is just sufficient to stop the Universe expanding at some infinitely distant time. All the visible matter in the Universe adds up to only a tiny fraction of the critical density. Can dark matter tip the scales? Or is the Universe dominated by something even more bizarre, such as the energy that is created by the vacuum of space that is forcing the Universe to expand forever against even the mighty pull of dark matter? If true, then the bulk of the Universe is truly dark.

It is ironic that as telescopes became larger, and their detectors more sensitive and wide ranging in their spectral reach, they revealed not a Universe filled with light, but one plunged into darkness; not a Universe dominated by blazing suns and galaxies, but one ruled by an invisible, as yet unidentified, substance. The stars and galaxies may sparkle like jewels, but in a sense that is only because they shine against the velvet blackness of dark matter. Despite their telescopes, their detectors, and their initial objections, astronomers have been forced to ponder a largely invisible Universe. Yet they continue to investigate dark matter through their telescopes, in their laboratories and with their theories. It is an intense search that is taxing some of the most brilliant minds the world has ever known, and occupies great slices of precious observing time on the world's most advanced telescopes. It seems strange to use telescopes to search for something invisible, something that emits no light. But just as such investigations revealed the outer members of our Solar System, so the search for dark matter will eventually reveal the rest of the Universe. It may have begun as a quest for light, but now astronomy is a quest for darkness.

1

How to weigh galaxies

There are no purely observational facts about the heavenly bodies. Astronomical measurements are, without exception, measurements of phenomena occurring in a terrestrial observatory or station; it is by theory that they are translated into knowledge of a Universe outside.
Arthur S. Eddington, *The Expanding Universe*, 1933

INTRODUCTION

The Universe seems to be dominated by dark matter. By studying the dynamics of visible matter – the movements of stars and galaxies – astronomers* have not only found that there are forms of matter other than that we can see, but that this luminous matter is actually in the minority, outweighed in some cases a hundred to one by dark matter. To say that it 'seems' to be dominated is scientific caution, as nothing is ever really proven in science. Nonetheless the evidence for dark matter is overwhelming. Using sophisticated techniques, astronomers are now able to study the kaleidoscopic phenomena of the Universe with increasing precision, and ever tinier movements of ever fainter objects are becoming observable. Time is routinely measured on scales from minutely split seconds to the very age of the Universe. The visible Universe has now been studied using almost the entire spectrum of electromagnetism, and at every turn, evidence for dark matter is revealed.

What is it, specifically, that suggests to us that dark matter exists at all, let alone in such vast quantities? The answer lies in the conflict between two measurements of the mass of the Universe: luminous mass and gravitational mass. In other words, there is a conflict between the total mass of all we see in the form of luminous stars and galaxies, and the mass implied by their motion through space which in turn implies a gravitating, although unseen, mass. These two concepts – luminous mass and gravitating mass – are central to the story of

---

* Throughout this book we will refer to astronomers who study dark matter, although those that study the problem now include physicists, astrophysicists and engineers. The subject is so interwoven within these fields that it is now impossible to distinguish them.

dark matter, and so we begin by talking about how astronomers weigh the Universe. (Moreover, a new spectrum, that of gravitational waves – ripples in the fabric of spacetime – may soon be opened for study.)

HOW TO WEIGH GALAXIES

The basic tool astronomers use to determine the mass of a system of stars or galaxies is to study their motion through space, and then compare that motion with the gravitational force needed to keep the system bound together. It was Newton who first showed that the motion of objects could be explained by the sum of various types of forces. When different forces act, the resultant motion is the sum of the effect of each different force. Especially in the case of gravity, we have a law which epitomises the concept of laws in physics: Newton's laws apply equally everywhere in the Universe. The balance between motion and gravity is often obvious and beautiful, perhaps best visualised by thinking in Newtonian terms of a balance of forces. We are surrounded by some wonderful examples, such as the Moon which silently orbits the Earth with mathematical precision simply because the gravitational attraction between the two bodies almost exactly balances the Moon's desire to keep moving in a straight line. Some examples are stunning; for example, the rings of Saturn, which are made up of countless particles. The rings display a symmetry so perfect that it is tempting to ask why the particles do not fly around the planet like a halo of moths around a streetlamp. Indeed, why do they not simply fly off into space, or plummet towards the planet? The solution is that many of the original particles that must have surrounded Saturn *did* fly off into space or become part of the planet, but they did it long ago. What we see today is all that remains – those particles that are trapped in a delicate balance between the forces of motion and gravitation. This same balance is repeated in the congregation of asteroids into the asteroid belt, or the spiral formation of stars and gas within the Milky Way.

NEWTONIAN GRAVITATION AND FINDING THE INVISIBLE

While it has been said that Newton's ideas on gravitation are simply an approximation, it is wrong to underestimate them. They have been good enough to reveal unseen masses at a variety of scales. In fact, the first ever experience with dark matter occurred in our own backyard, the outer Solar System. Despite its success at describing the motions of most of the known planets, for a while it seemed Newton's laws were failing with the seventh planet, Uranus, whose erratic wanderings refused to follow Newton's laws of gravity. No matter how many ways the celestial mechanicians manipulated the numbers, Uranus just would not follow its predicted path among the stars. Here was a problem. Could it be that Newton's gravitation had a limited range, and that beyond a certain distance from the Sun it broke down, allowing planets to wander unleashed throughout the starry sky? Perhaps that is an exaggeration, since the amount that Uranus' observed position differed from its predicted position amounted to the equivalent of the width of a human hair seen from a distance of a 100 metres! Yet this tiny amount annoyed the astronomers of the time like nothing else. What was causing the error?

Enter two bright, young men who each had a flare for mathematics. One was British, John Couch Adams; the other was a Frenchman, Urbain Jean Joseph Leverrier. By the early 1830s, the problem of Uranus' wanderings had become so pronounced that astronomers began to wonder whether it might be the presence of another planet still further from the Sun. Such a planet would exert a gravitational pull on Uranus, tugging it from its predicted location. Adams was the first to accurately estimate the unseen planet's mass, distance from the Sun, and, most importantly, location in the sky. By October 1843, Adams had a reasonable estimate of where in the sky an observer might find the new planet, but owing to petty snobbery and personal differences, he could not gain the interest of the Astronomer Royal, George Biddell Airy, and his predictions remained untested.

Two years later, on the other side of the English Channel, Leverrier performed similar calculations to Adams', quite unaware of the earlier results. Leverrier completed his work on 18 September 1846, and passed the results to the German astronomer Johann Gottfried Galle. Galle had, quite by chance, recently acquired a new set of star charts covering the area of sky containing the predicted position of the new planet. He began looking for the new planet, and found it only a few days later on 23 September 1846... within a degree of the position predicted by Leverrier!

Controversy raged over who should be given credit for the discovery of the new planet, later to be called Neptune. James Challis – Airy's successor as Professor of Astronomy at Cambridge – claimed he had found Neptune during his own searches but had not had time to verify his discovery, while it was Galle who had been the first to positively identify Neptune through a telescope. Ultimately, history has credited Adams and Leverrier jointly, although Adams received very little recognition in his own time. (This account was published in *Ripples on a Cosmic Sea* by David Blair and Geoff McNamara, Allen & Unwin, 1997.)

This was a major triumph for Newton's gravitation, as even the tiniest of discrepancies that led to the investigation in the first place were explained by the laws of gravity. Gravitation was universal, and the movements of more distant and more massive objects could therefore be studied to probe the distribution of matter in the wider Universe. But advances in understanding the Universe at ever larger scales needed more than simply a good theory of gravity; it also needed better observing techniques. Measuring the movements of stars is a much more difficult task than measuring the motions of planets – for a very simple reason – stars are much further away. Driving down the motorway you will have noticed that the nearer the scenery, the more it appears to move relative to yourself. As you look further towards the horizon, however, the trees and hills seem to move

Figure 1.1. M74 (NGC 628) – a spiral galaxy similar to the Milky Way. These beautiful stellar structures are the result of inward gravitational forces and the random and circular motions of the stars which keep the galaxies 'inflated'. Unlike elliptical galaxies, they are dominated by circular motion. (Courtesy Todd Boroson/NOAO/AURA/NSF.)

at a snail's pace no matter how fast you drive. This phenomenon is called perspective, and it is at its best in the sky. Despite the 3,500-km/h motion of the Moon, you would have to watch it for several minutes before seeing an appreciable motion against the background stars. The planets seem not to move at all during the night, taking days, weeks or months to move even a few degrees. To study the distribution of matter out among the stars, it is essential to measure their motions. However, perspective means that even the tiniest movement of a star may take years to measure. Nonetheless, astronomers managed to do it, and in quite an ingenious way.

HOW TO MEASURE STELLAR MOTIONS

The motion of a star through space is defined in three dimensions: one dimension is along the line of sight, while the other two are across the plane of the sky. The velocity of a star along the line of sight is measured using a process called spectroscopy. When starlight – whether it be from an individual star or from a galaxy – is passed through a glass prism or similar device, the light is dispersed; that is, broken up into its component colours. The same thing happens whenever sunlight passes through rain and a rainbow is formed. The band of bright colours – violet, blue, green, yellow, orange and red – is called a spectrum. If the spectrum of a star is produced well enough, you can see within it a series of dark lines resembling a bar code. Each of these spectral lines represents a specific chemical element within the star, and has a specific and known position within the spectrum. (This also means astronomers can identify the make-up of a star simply by reading the spectral lines.)

Light is made up of electromagnetic waves, and each colour has a specific wavelength, that is the distance between the peaks of two waves. The position of a spectral line is therefore described in terms of wavelength: those lines with shorter wavelengths are found closer to the blue end of the spectrum and those with longer wavelengths are closer to the red end. The important thing for this story is that the motion of a star along the observer's line of sight causes the apparent wavelengths of light to change. The change – which can be either an increase or a decrease in wavelength, depending on the direction of motion – is caused by the Doppler effect, the same phenomenon that causes the change in pitch of a police car's siren as it approaches and then (hopefully) recedes from you as you drive down the motorway. When the police car is approaching you, the siren's sound-waves are compressed, creating a higher pitch; when receding, the waves are stretched out and so sound lower in pitch. The Doppler effect applies just as well to starlight. If the star is approaching you, its light-waves will be compressed so that any spectral lines will appear to have a shorter wavelength. This causes them to shift towards the blue end of the spectrum. If the star is moving away from you, on the other hand, its light-waves will be stretched and the spectral lines will be shifted towards the red end of the spectrum. It is the blue or redshift of a star's spectral lines that reveals whether the star is approaching or receding from you, respectively.

The concept can be taken further, however. The rate at which a star is approaching or receding determines how far the spectral lines will be shifted. This means that the line-of-sight velocity of a star (usually called its radial velocity) can be measured with remarkable accuracy by looking at whereabouts the spectral lines lie in the visible spectrum. Even back in the 1920s and 1930s, when all this was done photographically, the precision obtained was remarkably high. Stars move around the Milky Way at velocities ranging up to a few hundred kilometres per second, and astronomers can easily measure these velocities to an accuracy of 1–2 kilometres per second.

In all of this, it is important to remember that estimating the mass of a galaxy or galaxy cluster is a statistical procedure, and so the more stars or galaxies you observe the better. As we will see when we look in on modern astronomers taking measurements of stellar and galactic velocities, in the later decades of the twentieth century this process was aided greatly by the use of optical fibres. Using these techniques, it is possible to measure hundreds of objects simultaneously. But keep in mind that the pioneers of the dark matter frontier were restricted to measuring the velocity of one star or galaxy at a time, using smaller telescopes and less sensitive detectors.

Returning to a star's motion, its radial motion either towards or away from us is only half the story. To determine a star's true motion through space we also need to measure its motion across our line of sight, in the plane of the sky. This transverse movement is known as the star's proper motion, which is measured in a more direct though (surprisingly) far less precise manner. First of all, you take an image of a star and measure its position. Then wait as long as you can – a year, five years, ten... the longer the better – then measure the star's position again and look for any movement relative to other stars or galaxies. It may sound simple, but is in fact a very tough business and less accurate than radial velocity measurements simply because of the difficulty of measuring such tiny motions across the sky. Astrometry from space is more accurate. A spectacularly successful satellite called Hipparcos has recently provided astronomers with detailed astrometric data on stars in the sky down to the twelfth magnitude (some 250 times fainter than the faintest star visible to the naked eye). The results confirm the absence of dark matter in the disk of the Milky Way. Astronomers are looking forward to the next astrometric space mission, the European satellite GAIA, which is scheduled for launch in 2011 and will provide astrometric data for about a billion stars down to twentieth magnitude.

At any rate, if you can acquire these two pieces of information – the radial velocity of a star and its transverse, or proper, motion – then you can work out its true three-dimensional motion through space, the so-called 'space motion' of the star. Although the techniques have been refined over the last hundred years or so, measuring the space motion of stars remains one of the most difficult tasks for astronomers.

HOW GALAXIES STAY INFLATED

The study of the motions of stars has played an essential role in unravelling the shape and determining the mass of the Milky Way we live in. Whenever you see a stellar system, you are always asking: 'Why does this thing not just collapse in on itself? What is holding it up?' A system of stars like the Milky Way stays inflated only because of the motions of the stars within it. This motion comes in two forms: average and random. If you were to measure the space motion of various stars passing through some region close to the Sun, you would notice that they all have an average motion that is mostly rotation – the Sun and the other stars are rotating around the Milky Way together. But it is not an absolutely smooth motion. The stars are not going around in perfect circles, they are going around in perturbed circles that resemble a badly buckled bicycle wheel. So a typical star that is going around the Milky Way at the same distance from the centre of the Milky Way as the Sun – say 8 kiloparsecs – has a motion that is roughly circular. (1 kiloparsec is equal to about 3,262 light-years.) As the star goes around the Milky Way, however, it wobbles around that circle –

sometimes closer to the Galactic centre, sometimes farther out – with an amplitude of 0.5–1 kiloparsec. These oscillations are quite random. If you were sitting near the Sun looking at those stars going by, you would notice that some of them are oscillating inwards at this point and some outward at this point, while others are at the turning point in their oscillation. Overall it is a sort of average circular motion, which is why the outline of the Milky Way is circular plus some random motion. Let us look at these two types of motion – circular and random – in more detail.

CIRCULAR MOTION

In pure rotation, everything goes around in a circle. This requires an inward acceleration (an inward gravitational pull), which is just the velocity squared divided by the radius. The inward gravitational pull depends on the position in the galaxy. If every particle in a galaxy is going around at just the right velocity for whatever radius it lies at, so that the gravitational field provides just enough acceleration to make it go around in a circle, then the system is in perfect circular motion. In such a system the random motions are zero.

RANDOM MOTION

At the other extreme is a completely random system in which all the stars are plunging in towards the centre of the galaxy and out the other side in random directions, and in which there is no average rotation at all. Such a system would include stars of different energies reaching out different distances from the centre of the galaxy, and the whole affair would be held up entirely by random motions of stars, with the random motions acting like the pressure in a gas. Such a galaxy would look like a swarm of moths around a street-lamp.

In a real galaxy, of course, it is a mixture of the two. Spiral galaxies like the Milky Way have mostly rotation with a little random motion thrown in, while elliptical galaxies have more random motion and less rotation. But at every point in every galaxy, three factors have to balance: the inward gravitational force and the outward pressure from the random motions of the stars must together provide the acceleration needed for the average circular motion. Each of these three factors changes with position in the Milky Way.

THE JEANS EQUATIONS

A set of equations that relates these three factors was formulated by the British scientist Sir James Jeans in 1919. We will learn more about Jeans' equations (a strictly non-mathematical explanation) in the next chapter, but for now it is worth mentioning that they relate mass and motion at a variety of scales, from

Figure 1.2. The giant elliptical galaxy NGC 1316, in the Fornax Cluster. Elliptical galaxies have much less structure than spiral galaxies because of their stars' random motions. Although invisible over human lifetimes, these stars are actually plunging through and around the galaxy like moths around a streetlamp. (Courtesy P. Goudfrooij (STScI), NASA, ESA and The Hubble Heritage Team (STScI/AURA).)

the interstellar molecular clouds in which stars are formed to the galaxies themselves. At the Galactic scale, Jeans' equations relate the density of stars at a point in the Milky Way (that is, the amount of matter in a given volume of space) to the average and random motions of stars and the gravitational force acting at that point.

By using Jeans' equations, astronomers have been able to use the observed motions of the stars to determine how much mass the Milky Way contains, implied by the motions of the stars. The mass is implied by the amount of gravity needed to balance the motions of the stars, and so is called gravitational mass. Now, the amount of gravitational mass is just fine… until compared with the amount of mass implied by the luminous matter contained within the Milky Way – the so-called luminous mass.

MASS–LUMINOSITY RELATIONSHIP

In the early decades of the twentieth century, determination of the amount of visible matter was a pretty indirect procedure. Astronomers had a vague idea of what kinds of stars are to be found in the Milky Way, and in 1924 the British astronomer Arthur Stanley Eddington predicted an approximate relationship between the mass of a star and its absolute brightness (the brightness a star would be if it were at a standard distance of 10 parsecs). Later on, this relationship was refined empirically from studies of binary stars. The mass of a star is usually given as a comparison with the mass of the Sun, in 'solar masses', where the Sun = 1. However, it should not be assumed that the relationship

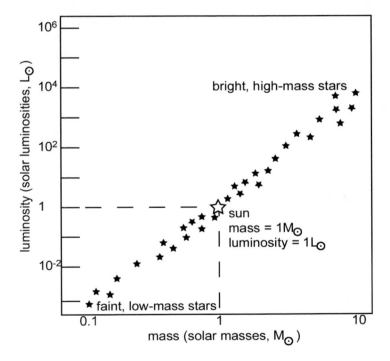

Figure 1.3. The relationship between the masses of stars and their brightness (luminosity) is illustrated in a mass–luminosity diagram. Understanding the mass–luminosity relationship allowed astronomers to calculate the Milky Way's 'luminous mass'; that is, its mass if it were to consist only of stars.

between absolute brightness and mass is the same for all stars. For example, a massive star is usually very bright, so the mass–luminosity ratio is very small. In other words, there is a lot of luminosity for a given amount of mass. At the other end of the scale, very small stars – much smaller than the Sun – do not shine very brightly at all compared with their mass. Here the mass–luminosity ratio is very high. The mass–luminosity relationship for stars of different brightness has been studied carefully over the years, and is now fairly well understood for the vast majority of stars in the Milky Way. Therefore, by adding the masses of all the stars of different brightness in the proportions seen in the Milky Way, it is possible to make an intelligent estimate of the Milky Way's 'luminous mass'. This was originally done in the 1920s, and although more modern estimates are produced in a more theoretical way (such as by making populations of virtual stars on a computer), the answer comes out pretty much the same.

GRAVITATIONAL VERSUS LUMINOUS MASS

Now comes the dilemma facing astronomers who study the dark matter problem. If the stars and gas in the Milky Way were all that existed, then the gravitational mass and the luminous mass of the Milky Way would be the same. But as you might have guessed, they are very different. Just how this was discovered, and by whom, is the subject of the next chapter in our story.

2

The false dawn

HISTORICAL BACKGROUND

To tell the story of how observations of stellar motions led to the discovery of dark matter we have to go back in time. But not too far back – just the 1920s, in fact, when our awareness of the scale, nature and origin of the Universe was expanding at an unprecedented pace. This period in history is comparable with the early days of the telescope when the Universe was transformed from a parochial, geocentric structure to one in which the Earth was nothing more than another planet orbiting an ordinary star. By the 1920s many astronomers accepted the model of the Galaxy proposed by Jacobus Cornelius Kapteyn in 1901: an immense elliptical congregation of stars, with our Solar System somewhere near the centre. Although he did not publish his theory for another twenty years, Kapteyn's model was well known amongst his colleagues, including the young Jan Hendrik Oort, who we will meet shortly.

Despite the popularity of Kapteyn's model, the work of many astronomers, including the Americans Harlow Shapley and Heber D. Curtis, showed that not only was the Milky Way much larger than that envisioned by Kapteyn, but the strange, spiral-shaped nebulae scattered across the sky and assumed to be part of the Milky Way were in fact galaxies of immense proportions lying at fantastic distances from our own. Further observations using techniques similar to the spectral line observations described in the last chapter were made by astronomers including Vesto Melvin Slipher and, later, Edwin Hubble. By far the majority of galaxies studied showed a redshift, indicating that they are moving away from our own Milky Way. But rather than implying that the Milky Way is at the centre of the Universe, this discovery showed that the Universe is expanding, and that each galaxy is receding from every other galaxy. This mutual recession of the galaxies is often illustrated by imagining sultanas in a fruit cake cooking in an oven. As the cake rises, each sultana moves away from the others, carried by the expanding cake.

Although Hubble identified this universal expansion, we should note that his observations of receding galaxies were preceded by Slipher, who at that time was the leading authority on galactic velocities. What Hubble did was to discover a simple law that relates distance to recession velocity. An important implication of the mutual recession of galaxies is this: the faster a galaxy is receding from you, the further away it is. For example, imagine three galaxies equally separated along a line and all moving away from each other at the same rate. Now imagine us living in a galaxy at one end of the line. From our perspective, the middle galaxy will be moving away from us just as we would appear to be moving away from inhabitants of the middle galaxy. But if we were to measure the recession of the second galaxy (which is twice as far away) it would appear to be moving away twice as fast as the nearer galaxy. This is because it is receding from the nearer galaxy, which is itself receding from us. In other words, the greater the recession velocity of a galaxy, the further away it is. Because the recession velocity is proportional to the distance, the ratio of the velocity to the distance is approximately constant as we go from galaxy to galaxy. This is called the Hubble constant, and it has also been notoriously difficult to determine accurately. One reason is that it is so difficult to measure precise distances to galaxies.

Hubble paid little attention to the theory behind just why the Universe might be expanding. As Hubble and Slipher studied the receding galaxies, it was left to several theorists – including Albert Einstein – who were developing a theoretical framework that involved such an expanding Universe. An important implication of an expanding Universe was the beginning of all things at some finite time in the past. This event has come to be known as the Big Bang, and is crucial to our story on dark matter. We will explore its intricacies in more detail later.

INTRODUCING OORT

Into this era of exploration strode one of the giants of twentieth-century astronomy, Jan Hendrik Oort. Although his own conclusions about dark matter are now believed to be incorrect, it was Oort, more than any other, who inspired astronomers thinking seriously about dark matter on galactic scales. Although not physically large, Oort was a very heavy-duty individual with a powerful presence. During the Second World War, he refused to cooperate with the German authorities and left his position at Leiden University to continue his research from a cottage in the Dutch countryside. Oort was very respected internationally from an early age, and carried a lot of political weight in the Netherlands. After retirement, he pursued his research almost to the time of his death at the age of 92, and was still ice-skating well into his eighties.

The Netherlands has a long and strong tradition of astronomy, and Oort continued (and led) this tradition. He became Professor of Astronomy at Leiden Observatory in 1935, around the time that radio astronomy was emerging in Australia and the United States. He became increasingly interested in radio astronomy, realising that not only would radio waves penetrate the obscuring gas and dust clouds of the Galaxy, but also the terrestrial clouds which dominate Dutch skies on most nights. It is therefore no surprise that at the end of the Second World War he and his colleagues obtained a German 7.5-metre radar

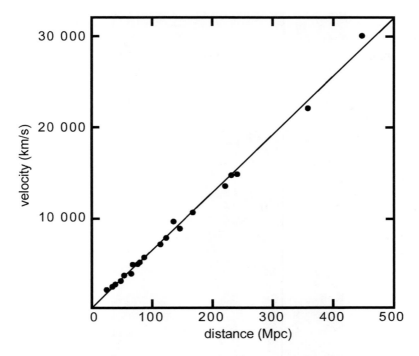

Figure 2.1. The redshift–distance diagram, showing how the velocity of recession of galaxies increases with their distance. It is based on observations of Type 1a supernovae, which all have about the same brightness and can be used to measure the distances of their parent galaxy quite accurately. The distances are given in Mpc (1 Mpc = 3 million light-years). The data derive from work by Riess and colleagues in 1996, and the figure is taken from Edward Wright's web page (www.astro.ucla.edu/~wright).

dish, converted it into a radio telescope, and used it to carry out ground-breaking research. It was one of Oort's students, Henk van de Hulst, who predicted the emission of the now famous 21-cm spectral line of neutral hydrogen. The Leiden astronomers were among the first to observe the motion of the hydrogen gas that lies between the stars in the Milky Way, enabling them to study its rotation. These studies – also carried out by astronomers in the United States and Australia – confirmed the long-suspected spiral structure of the Milky Way. These were history-making observations.

Oort realised, however, that the Netherlands needed a modern radio telescope much larger than the 7.5-metre German antenna that they had been using. In 1956 he was instrumental in the construction of a 25-metre antenna at Dwingeloo in the north of the Netherlands, and in 1970 the Westerbork Radio Synthesis telescope. The Westerbork instrument is a giant interferometer consisting of fourteen 25-metre dishes along a 2.8-kilometre east–west track, and is similar to the Australia Telescope Compact Array (ATCA), though larger

Figure 2.2. Jan Oort at the International Astronomical Union conference in Brighton, England, in 1970.

and more powerful. As we will see in Chapter 4, the Westerbork telescope was very important in making astronomers realise that spiral galaxies consist mostly of dark matter, not stars. In short, Oort's influence in this field was very great. But we should return to the story of his earlier research.

OORT DISCOVERS DIFFERENTIAL ROTATION

One of the first major discoveries made by studying the space motions of stars was the fact that the Milky Way rotates. There were three major figures involved in the discovery of the Galaxy's rotation that broke between about 1925 and 1930: Gustav Stromberg, Bertil Lindblad and Jan Oort. Stromberg's contribution was the discovery that the average motion of Milky Way stars correlates with their random motions: the bigger the random motion of some particular class of star, the bigger its average motion relative to the Sun. The random motions of the stars act like gas pressure, and together with the rotation they help to support the galaxy against its own gravitation. Stromberg's discovery was a major clue that we live in a rotating galaxy. Lindblad was the first to show that the Milky

Way was actually rotating, with stars following nested orbits about the centre of the Galaxy, much like the planets orbits around the Sun, but more complicated. Lindblad's mathematical papers inspired many astronomers, including Oort, to study the motion of stars within the Galaxy.

Oort's contribution was his revelation, in 1927, of the Galaxy's differential rotation. He realised that the disk-shaped Galaxy did not rotate rigidly like a car's wheel where every part is moving around with the same period, but rather the inner parts of the Galaxy were spinning faster than the outer regions. Every point in the Galaxy rotates with a different angular velocity, so that the motion of the stars and gas in the Milky Way resembles the vortex produced as bath water goes down a plug-hole. As simple as it sounds, this was a major achievement at a time when the structure of the Galaxy – that is, a spiral-shaped disk – had not yet been confirmed.

OORT 'DISCOVERS' DISK DARK MATTER

Oort's interest in the dynamics of galaxies subsequently led him, in 1932, to a detailed study of the amount of matter in the disk of the Galaxy. Both Kapteyn and Jeans had already expressed suspicion of the existence of invisible matter in the Milky Way, but it was Oort who carried out the first definitive study of the problem. To measure the amount of matter in the disk of the Milky Way, Oort examined not just the average and random motions of stars around the Galaxy, but also their motion perpendicular to the plane of the Galaxy. Stars go around the disk, but like toy horses on a merry-go-round they are also moving up and down through the plane of the Galaxy. In the same way that the average (rotational) and random motions of stars prevent them from collapsing towards the centre of the Galaxy, it is their vertical motion that prevents the Galaxy from collapsing to a structure resembling a completely flat disk. Now, just as one of the Jeans equations relates average rotational motion, random motion and gravitational force within the plane of the Galaxy, another Jeans equation relates the vertical motion of stars to the gravitational force in the disk. The gravitational force that pulls all the stars down towards the plane of the Galaxy is balanced by the random vertical motions of the stars, which again act like the pressure in a gas.

When Oort believed he had discovered dark matter in the disk, he was looking at these vertical stellar motions; but this is not easy, as much can go wrong. And in Oort's case, something *did* go wrong. The observable factors that enter into the measurement of the density of matter in the disk are the distribution of stars perpendicular to the disk and their random vertical motions. The Jeans equations are, however, very fussy about which stars are used for the study. Exactly the same stars must be used to acquire the density and motion data, otherwise the results will be erroneous. This is where Oort went wrong.

To measure the vertical distribution of the stars perpendicular to the disk of the Galaxy, astronomers of the time simply counted all the stars they could see in a defined area of sky that represented a column that was vertical to the disk. By comparing the distribution of the brightness of those stars, they deduced how the stars were physically distributed within the column of space. Statistically, seeing more bright stars than faint stars indicates that there are more stars close to you than farther away. Alternatively, fewer bright stars and fewer faint stars, but plenty of intermediate brightness stars, suggests a layer of stars at some intermediate distance, and so on. Of course, this ignores some important factors, not the least of which is the fact that not all stars have the same intrinsic brightness. Nor does it account for interstellar dimming by dust along the line of sight.

THE PROBLEM WITH K STARS

The vertical motions of the stars were measured using the Doppler effect. In the 1920s and 1930s, astronomers did not have the sophisticated detectors and spectrographs, much less the huge telescopes available today, and so they chose whatever stars were easiest to measure. There is a particular type of star that is very attractive to astronomers when making velocity measurements: K giants.

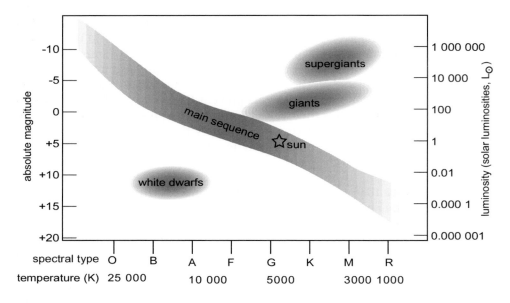

Figure 2.3. The Hertzsprung–Russell diagram plots the brightness of stars against their temperature (determined from their spectral type). It was independently devised by the Danish astronomer Ejnar Hertzsprung (1911) and the American Henry Norris Russell (1913). Any large group of stars – such as a cluster or galaxy – will form well-defined groups depending on the evolutionary stage of the stars.

(The letter K comes from the Hertzsprung–Russell diagram – a graph that compares the brightness and spectral temperature of a star.) These are older stars that have used up the bulk of their store of hydrogen, and have swelled to enormous proportions, typically thirty times the diameter of the Sun, and are consequently a hundred times brighter and hence easier to see. Just as importantly, they are cool stars, and therefore have many easily identifiable spectral lines to measure. (Hot stars, as a rule, have fewer lines for reasons that are not important here.) Oort assumed – as did all astronomers up to the 1980s – that these K giants were all typical members of the Galactic disk. But they are not.

THIN DISK AND THICK DISK

The disk of the Galaxy is quite complicated. As well as the thin disk portrayed in images of edge-on spiral galaxies, there is a rather diffuse feature called the thick disk, which was not discovered until 1983. The thick disk is just that: a disk about four to five times thicker than the thin disk with which most people are familiar, and replete with old stars, including the easy-to-measure K giants. The stars in the thick disk probably originated during the very early days of the Galaxy. One of the factors that complicated Oort's calculations was that the stars that occupy the thick disk have typically twice the random velocity of the stars in the thin disk (which is why the thick disk is thicker than the thin disk). These energetic thick-disk stars masqueraded as thin-disk stars and worked their way into Oort's calculations. Their greater vertical motions undoubtedly contributed to his conclusion that the average vertical motion of the stars in the disk of the Galaxy was a lot higher than it really is. Based on such a high average vertical motion, Oort concluded that there had to be much more matter – perhaps twice as much as was represented by the stars of the Milky Way – to hold the disk together. Such matter, if it existed, was completely dark.

BAHCALL AND THE RESURGENCE OF INTEREST IN DISK DARK MATTER

Oort continued his studies of disk dark matter through the 1940s and 1950s, but surprisingly the subject lay relatively dormant from then until the 1970s when an American astronomer, John Bahcall, who runs the astrophysics group at the Institute for Advanced Study in Princeton, carried out a similar exercise to Oort's but with more modern data. Bahcall's results were more or less the same as Oort's, but in contrast they sparked several other investigations into the idea of disk dark matter. By this time the idea of dark matter in the outer parts of galaxies had become generally accepted, so there was much interest in discovering whether dark matter was also really present in the disks of galaxies.

OORT'S ERROR REVEALED

This was the beginning of the end for disk dark matter. The ensuing independent studies consistently produced disk dark matter values smaller than either Oort's or Bahcall's. A major blow to disk dark matter came in 1988 when Konrad Kuijken – a Belgian student working with Gerry Gilmore at Cambridge – published an influential series of papers in *Monthly Notices of the Royal Astronomical Society*, suggesting that there was little, if any, dark matter in the disk of the Milky Way.

The different conclusion about disk dark matter is partly due to an improved mathematical method for measuring the amount of gravitational mass in the Galaxy's disk. The Jeans equations, as used by the earlier researchers, derive from a deeper underlying equation called the Boltzman equation, which relates not just how things happen in space but also in velocity space. Instead of saying at some point in space we have so many solar masses per cubic parsec – three-dimensional space – the Boltzman equation uses a density D defined in six-dimensional space. What this means is this. We sit at some place in the Galaxy and ask: How many solar masses per cubic parsec are there? But then we also ask about the density in velocity space: How many solar masses per cubic parsec per cubic kilometre per second? This is called the phase space density, usually denoted by f, and is a measure of how many stars of a given velocity exist in a given volume of space. The Boltzman equation determines how that density changes as a function of velocity and of position, and it therefore depends on the gravitational force. It is a rather more basic equation because, if the phase space density is known, a complete description of the stellar system is produced. While this technique has its own problems it is a more direct method, as it by-passes some of the variables required by the Jeans equations that cannot be measured very accurately.

Now, most astronomers believe that the disk of the Milky Way – contrary to Oort's conclusion – is more or less free of dark matter. In 1998 this conclusion received a strong boost from a study by Olivier Bienayme and colleagues in France. They used stars whose densities and motions had been measured by the Hipparcos satellite to show that the gravitating density of the disk is very close to the density that can be accounted for from visible objects.

NOT THE END OF DISK DARK MATTER

It must be stressed, however, that this does not necessarily mean the end of disk dark matter; only that the current thinking is that there is little if any dark matter in the disk. In fact it would make astronomical life easier if the disk *did* contain dark matter, as it would explain a number of observed phenomena. For example, as things stand now, the lack of disk dark matter means that the disk of the Milky Way appears to be quite a bit lighter than the disk of most otherwise similar galaxies. This is a festering worry. Is our Galaxy a little odd, or are some of the procedures we have used for measuring other galaxies wrong? Many astronomers find the absence of dark matter in the disk quite uncomfortable, and would prefer it to have about twice as much matter as it presently seems to have. Despite being apparently wrong, Oort's 'discovery' of disk dark matter would in fact be a much more comfortable result. There are other, smaller confusing factors that come into play, and that is why astronomers are not absolutely certain that even the present conclusions – that there is no dark matter in the disk – are correct, even though most people believe this to be the case.

But for two reasons, Oort's erroneous suspicion of disk dark matter was not simply a false trail. First of all, it helped refine the present model of the Galaxy. In 1922 Oort studied the way in which large, spherical collections of stars, called globular clusters, move within the Galaxy. Globular clusters are compact clusters of about a million stars which can be observed in the outer regions of most galaxies, including the Milky Way. Oort found that globular clusters move too fast to be bound to the Galaxy by the gravitational influence of the luminous matter alone. A simple star-chart shows that globular clusters seem to huddle about the centre of the Galaxy, in the direction of the constellation Sagittarius. Such an asymmetrical distribution in the sky strongly suggested to Oort that the globular clusters are part of the Galactic system, and not simply passing through. But in order to be able to retain them while they travelled at their observed velocities, the Galaxy would have to be 200 times more massive than the stars it contained, according to Kapteyn's model. This sparked not only interest in the idea of invisible (dark) matter, but also a rethink about the size and structure of the Milky Way.

Secondly, there is a lesson to be learned from the episode about the nature and process of science. It was several decades following Oort's announcement that half of the Galaxy's disk was made of dark matter before astronomers decided to investigate the problem further. It might be thought that such an important discovery, made by such an important astronomer, would have been taken very seriously by the astronomical community and investigated further, and we can only guess why it was not followed up at the time. Part of the reason was probably that the idea of dark matter was not part of everyday astronomical thinking at that time. But it seems that Oort's conclusion was taken a little too seriously. He was such a dominant figure that when he announced this result astronomers thought they could not do any better! It was not until Bahcall revived the problem that Oort's error was discovered. The lesson here surely is not to place too much faith in any one result, no matter how eminent the scientist who derives it. Science thrives and grows on doubt and scepticism. Without these, it stagnates.

Despite being a false start, Oort's work and his authority as a great scientist sparked the idea of dark matter on a galactic scale. Not long after he announced the existence of dark matter in the disk of the Galaxy, one of his contemporaries announced its existence on an immensely larger scale, and in far greater quantities. This contrast with Oort's findings is matched only by the contrast between the two men themselves.

3

Seeing the invisible

INTRODUCING ZWICKY

Soon after Oort had announced the existence of dark matter in the disk of the Milky Way, another astronomer, Fritz Zwicky, announced its existence not just in individual galaxies but in clusters of galaxies. At a time when few astronomers appreciated that our Galaxy was a fairly ordinary collection of stars no different from other galaxies in the Universe, Zwicky was already embracing a Universe not only with thousands of galaxies, but one with large-scale structure, and full of exotic phenomena such as neutron stars and dark matter.

Zwicky was an amazing scientist who made many important contributions, not only to astronomy but also in a wide range of other disciplines. Born in Bulgaria of Swiss parents, he spent most of his working life in America. After receiving a PhD in physics from the Swiss Federal Institute of Technology, Zurich, in 1922, he moved to the United States, where he served on the faculty of the California Institute of Technology, Pasadena, from 1925 until 1972. Despite his residence, in 1949 he was denied American citizenship.

Zwicky's astronomical contributions went further than just the discovery of dark matter on large scales. In collaboration with Walter Baade, in 1934 he pointed out the difference between novae and supernovae. They showed that supernovae occur much less frequently than novae, and correctly suggested that supernovae were the result of the death of massive stars. The remains of a supernova explosion, they said, would be a star made entirely of neutrons. As we describe in more detail in Appendix 1, the neutron is a small, chargeless particle found in the nuclei of most elements. Discovered in 1932 by James Chadwick, it is said that within hours of hearing of the discovery the great Russian physicist Lev Landau conceived of an unimaginably small and dense star composed predominantly of neutrons. But it was Baade and Zwicky who two years later described in detail how an ordinary star could turn into a neutron star. With some adjustments, the story they told remains basically the same today.

All stars are spheres of hot gas. Although they are immense, their spherical form is maintained against the pull of gravity by the constant flow of energy outward from the centre of the star. This energy comes from the fusion of lighter elements into heavier elements. So long as the star has a supply of nuclear fuel – such as hydrogen in the case of the Sun – it can maintain its opposition to the inward crush of gravity. However, all stars eventually run out of nuclear fuel, and the outward flow of energy comes to an end. A small star the mass of the Sun dies a relatively peaceful death. With nothing left to support it, its core collapses into a white dwarf about the size of the Earth, while the outer layers are blown off into space to form a beautiful 'planetary nebula'. A more massive star, however, ends its life in a much more spectacular fashion. Having used up its supply of nuclear fuel, it collapses in on itself violently. The tremendous implosion results in a 'core bounce' whereby the bulk of the star's outer layers explode with the light of a billion stars. Meanwhile, the stellar core continues to collapse. At least as massive as the Sun, the core would shrink to only about 20 kilometres in diameter. At this density, the electrons and protons would fuse to form neutrons... and a thimble-full of the material would weigh 100 million tonnes. It would be another thirty-five years before Zwicky and Baade's prediction would be confirmed.

Zwicky also predicted how dark matter in the Universe can be mapped using a phenomenon known as gravitational lensing (which will be covered in detail later in this book). The basic idea is that the gravitational fields of massive objects can deflect starlight. The concept is based on a prediction of Einstein's General Relativity: light follows the local curvature of spacetime surrounding an object with mass. The great British astronomer Arthur S. Eddington confirmed Einstein's prediction during a famous expedition to observe the bending of starlight during the total solar eclipse of 1919. Although the expedition was a success, and the confirmation of Einstein's prediction elevated him to unexpected fame, both Einstein and Eddington had doubts as to the practical application of the phenomenon to scientific research. Among the possibilities, however, was the concept of 'gravitational lensing', whereby the gravitational field surrounding a galaxy would bend – in fact, magnify – the light of more distant galaxies just as a magnifying lens concentrates the light of the Sun – a phenomenon well known to children at the expense of innocent insects! Despite the pessimism of Einstein and Eddington, who believed the chances of alignment were too small to be of serious concern, Zwicky predicted that gravitational lensing would be a useful tool for examining objects at the farthest reaches of the Universe. His prediction could not have been more profound. As we will see in Chapter 7, gravitational lensing is now used extensively in the search for dark matter not only within the Galaxy but also in galaxy clusters, with startling success.

As if these were not enough, Zwicky also contributed to research ranging from the study of cosmic rays to developing some of the first jet engines. Not many scientists are intellectually decades ahead of their time, but certainly Fritz Zwicky was a remarkable example of such a person. At the personal level he was a loud character with a very strong accent, despite living in the United States for almost five decades. He was bombastic and rather self-opinionated, and had a low opinion of many of his colleagues. One of his favourite insults was to refer to

Figure 3.1. Fritz Zwicky talks galaxies across a luncheon table at Prague, during the International Astronomical Union's gathering in August 1967. (Courtesy Sky Publishing Corporation.)

people he did not approve of as 'spherical bastards', because, he explained, they were bastards no matter which way you looked at them.

Above all, Zwicky was an observational astronomer, although he also carried out important theoretical work. He was one of the first to show that the apparent concentrations of galaxies – first pointed out by Herschel – were in fact true clusters. Using the 48-inch Schmidt telescope on Mount Palomar Observatory he discovered many galaxy clusters. As another example, at a time when the total number of supernova discoveries was twelve, between 1937 and 1941 Zwicky discovered a further eighteen in other galaxies.

Later in his life he produced some large catalogues of galaxies, but even here his acrimony emerged. As he became more bitter about his treatment by the outside world, he wrote a vitriolic foreword to his 1971 *Catalogue of Selected Compact Galaxies and of Post-Eruptive Galaxies*, parts of which make entertaining reading. Here are some examples.

> The naivety of some of the theoreticians, at all times, is really appalling. As a shining example of a most deluded individual, we need

only quote the high pope of American Astronomy, one Henry Norris Russell... who in 1927 announced 'The characteristics of the stars depend on the simplest and most fundamental laws of nature, and even with our present knowledge might have been predicted from general principles if we had never seen a star.'

Henry Norris Russell was indeed a very famous astronomer who made giant advances in stellar astronomy, and most modern astronomers would not take much issue with his announcement!

Zwicky was not permitted to use the Mount Palomar 200-inch telescope until late in his career, which clearly riled him. He wrote:

The most renowned observational astronomers in the 1930s also made claims that have been proved to be completely erroneous... E.P. Hubble, W. Baade and the sycophants among their young assistants were thus in a position to doctor their observational data, to hide their shortcomings and to make the majority of the astronomers accept and believe in some of their most prejudicial and erroneous presentations and interpretations of facts. Thus it was the fate of astronomy, as of so many other disciplines and projects of man, to be again and again thrown for a loop by some moguls of the respective hierarchies. To this the useless trash in the bulging astronomical journals furnishes vivid testimony.

Zwicky was certainly consistent in choosing distinguished targets for his vitriol. Hubble and Baade were among the most influential astronomers of their time, and there is no doubting the importance of their discoveries about the expansion of the Universe and the kinds of stars that inhabit galaxies.

In sharp contrast to their ready and uncritical acceptance of all sorts of childish phantasies [sic] and stolen ideas, the Editors of the *Astrophysical Journal* exhibited an almost unbelievable lack of tolerance and good judgement by rejecting my first comprehensive and observationally well documented article on compact galaxies.

This was around the time of the discovery of quasars – immensely bright and distant objects.

Zwicky's contempt of his colleagues was sometimes reciprocated. He wrote a book called *Morphological Astronomy* in which he described 'morphological arguments' for using patterns of things as a way of determining their nature. The book is a rambling discourse on many of the views that he had developed over the years. However, it contains some glimmers of truth that remain relevant today. For example, it includes some of the first examples of composite imaging, in which one superimposes a negative print of a galaxy in one wavelength band (for example, yellow light) on a positive print in another wavelength (such as blue) to show how stars of different kinds are distributed within the galaxy. Nevertheless, most people disregarded this book. There is a story that at one observatory library Zwicky used to visit, his book was stored in the Fiction section, except when he was there... when it was relocated under 'Z' in the Science section.

Whatever else, it must be said that Zwicky was not afraid to address the big questions in astronomy, and he was quite clearly interested in figuring out how to determine the mass both of the extragalactic nebulae (the galaxies) and of clusters of galaxies. His first papers on these subjects were published in German around 1933, but in 1937 he wrote a detailed and substantial paper for the *Astrophysical Journal* outlining the basics of all the work on the measurement of mass that was to follow over the next sixty years, including matters such as the 'virial theorem' (a method for calculating the mass of a galaxy or galaxy cluster, which we will shortly describe). He also described the technique he used to discover dark matter in clusters of galaxies. Just as we can estimate the mass of an individual galaxy from the motions of its gas and stars, so can we measure the mass of clusters of galaxies from the motion of their individual galaxies. Zwicky's work on galaxy clusters was to open up an entirely new way of measuring mass in the Universe.

GALAXY CLUSTERS

In order to study the behaviour of galaxies in groups, it pays to have a large sample, and fortunately there is a number of galaxy clusters that are almost tailor-made for the job. The history of galaxy clusters originates with William Herschel who, in the 1780s, noted that galaxies (or 'nebulae', as they were then known) were not randomly distributed across the sky, but appeared to congregate in groups. One such gathering noted by Herschel was in the constellation of Virgo. Beginning in 1933, Harlow Shapley (who was instrumental in revealing the true size of the Milky Way through studies of Cepheid variable stars in the globular clusters) published a catalogue of twenty-five galaxy clusters and suggested that the congregations were not due to chance, but to some evolutionary process. Yet until the 1950s, most astronomers considered galaxies to be loners, with only a few being associated with others in any true physical connection.

ZWICKY AND ABELL CLUSTER CATALOGUES

Evidence that galaxies were gregarious began to accrue, however. In 1958 George Abell published a catalogue of 2,712 galaxy clusters, and between 1960 and 1968 Zwicky and co-workers published catalogues of 9,134 galaxy clusters. Both of these monumental works were created by scrutinising the photographic plates taken with the 48-inch Schmidt at Palomar, which had been compiled into the National Geographic Society–Palomar Observatory Sky Survey (POSS). The huge difference in the number of galaxies relates to the difference in the way the two astronomers categorised their 'clusters'. Abell used only the densest clusters, while Zwicky used any aggregate, no matter how sparse. Since then, increasingly sophisticated observing techniques and statistical methods have proven that the vast majority of galaxies belong to groups and clusters, and only a few are truly alone in the Universe.

Galaxies come in an array of shapes and sizes. As we have seen, we live in a flattened disk with an internal spiral structure. Such 'spiral galaxies' are usually rich in gas and dust, as well as in stars, and come in two forms: pure spirals that resemble cream spiralling on the surface of a freshly stirred cup of coffee, and spirals with bars across their centres from which the spiral arms extend. Elliptical galaxies, in contrast, have little internal structure and not much gas or dust, and are shaped like flattened beach-balls. Finally, there are the gas-rich galaxies, which have no discernible shape at all and so are called 'irregular'. The Small Magellanic Cloud is of this type. In 1925 Hubble presented a scheme of linking the classifications of galaxies. Although there is still confusion over the origin of the different galaxy types, recent observations of more and more distant galaxies – those that existed closer to the beginning of the Universe – are adding support to the link between morphology and evolution of galaxies.

Just as individual galaxies can be classified according to their appearance, galaxy clusters have identifiable characteristics. Abell and Zwicky were the first to attempt to identify cluster types. Abell defined them either as regular or irregular, while Zwicky defined them as either compact, medium or open. Neither classification scheme, however, took account of the types of galaxies within clusters, being based solely on the spatial distribution of the member galaxies. It was William W. Morgan who first examined the nature of the individual clusters. He studied a small number of nearby galaxy clusters in detail, concentrating on the brighter galaxy members (within two magnitudes of the brightest member), and identified two types: those that are rather spread out and contain mainly spiral galaxies and where only the very brightest are elliptical or S0 (flat like spirals, but containing little gas); and denser clusters in which all the brightest members are either elliptical or S0 galaxies, with hardly any spirals other than the fainter members. As with individual galaxies, these characteristics suggested a link between the morphology of a cluster and its environment.

Galaxy clusters are now defined as either regular or irregular. Regular clusters are roughly spherical aggregations of galaxies that contain mainly elliptical or S0 galaxies. They are rich and dense, and may harbour 5–10% of the galaxies in the Universe. Irregular clusters, on the other hand, have no discernible shape and can contain any type of galaxy. These clusters are clumpy, resembling groups of smaller clusters, and can contain from ten to 1,000 galaxies, although smaller clusters (with diameters of about 3 million light-years) are more common. Literally thousands of clusters of galaxies have been discovered, and their study has provided astronomers with insights into disciplines ranging from galaxy formation and evolution to high-energy astrophysics and large-scale structure and cosmology.

THE COMA CLUSTER

The cluster that Zwicky used for his investigations into dark matter is listed in Abell's catalogue as 1656, and can be found about 90 megaparsecs away in the northern constellation of Coma Berenices. More commonly referred to as the Coma Cluster, this huge cluster – one of the richest known – was first described by Max Wolf, using the Bruce telescope in Heidelberg, Germany, in 1901. Its roughly spherical shape is divided into two distinct components, with each clump more or less equal in size and centred on a galaxy which is in turn associated with a radio source. The Coma Cluster is an ideal subject for observation. It is in a region of sky well away from the plane of the Milky Way, and so is not obscured by stars, gas and dust. In 1957 Zwicky found 804 galaxies brighter than magnitude 16.5 within about 2.5 from the centre of the cluster, and 29,951 galaxies brighter than magnitude 19 within 6. Many of these galaxies are unrelated to the Coma Cluster, but the system itself almost certainly consists of well over 1,000 galaxies, and so is one of fewer than 5% of the clusters

Figure 3.2. The Coma Cluster of galaxies lies at a distance of about 90 megaparsecs in the northern constellation of Coma Berenices. Consisting of thousands of individual galaxies, it is divided into two clumps, each centred on a dominant galaxy: the elliptical NGC 4889 and the SO spiral NGC 4874, which are both associated with radio sources. (Courtesy NASA.)

in Abell's catalogue that have as many galaxies. The galaxies that make up the Coma Cluster are mostly elliptical and S0 galaxies, and it has been estimated that they were among the first to have formed in the early Universe. Dominating the cluster are two giant elliptical galaxies, NGC 4889 and NGC 4874, which presumably have achieved their great size by absorbing or cannibalising smaller galaxies over their immense lifetimes.

MEASURING CLUSTER 'PRESSURE'

We saw in the last chapter how Oort and others have measured the mass of the Milky Way by measuring the motion of stars within it. Despite the immense difference in size and mass, the mass of a cluster of galaxies is measured in a similar way. One big difference is that in a galaxy cluster the main motion is not rotation but the random motion of all the galaxies. Fortunately there is a variety of methods for dealing with these motions. The behaviour of a cluster resembles the gas pressure in an inflated balloon, which remains inflated because the inward tension of the rubber is counteracted by the random motions or pressure of the air molecules inside it. In a similar way, an elliptical galaxy or a galaxy cluster remains inflated by the 'pressure' of the motions of its constituents – stars in a galaxy, or galaxies in a cluster – that resist the inward pull of gravity. The 'pressure' can be calculated by measuring how fast galaxies are moving in a cluster, and from this pressure it can be determined what gravitational field, and hence mass, is needed to contain the galaxies.

Now this is rather tricky, as some assumptions have to be made about the degree of random motion. For example, if every galaxy was simply moving radially in and out through the centre of the cluster, one value for the mass would be derived. But that would be a very unusual situation, and in fact most galaxies do not simply move in a radial orbit straight through the centre of the cluster. Some are moving in circles, and some are in radial orbits, but the majority pass through the cluster via every possible combination between these two extremes. It therefore needs to be assumed that there is a random combination of orbit shapes.

Yet another factor that enters into this equation is that systems like clouds of gas or clusters of galaxies are not simply a balance between gravity and pressure, but a balance between gravity and the gradient of the pressure; that is, how the pressure is changing with distance from the centre of the system. If the pressure is the same everywhere throughout the system, there are no pressure forces at play. However, if there is a higher pressure in the middle than at the outer regions, then there is a pressure force pushing outward. A good example of this is a single star, in which the pressure in the centre of the star is much higher than in the outer layers. The pressure gradient of the gas is pushing the star's matter outward, trying to make it expand, but the gravity is holding it back. Furthermore, when a pressure gradient is measured, the distribution of all the components of the system – whether they be molecules of gas, or galaxies in a cluster – is important. Although galaxy clusters are poorly defined, it is possible to define an average radius for a cluster mathematically, and therefore to determine the pressure gradient for the cluster.

To determine the random motions within the Coma Cluster, Zwicky studied the motions of about 600 member galaxies. Because of their immense distance it is impossible to detect any motion of the galaxies in the cluster across our line of site, as the perspective at these distances is extreme. However, by studying the Doppler shift of the spectral lines of individual galaxies he was able to measure the line-of-sight motions of the individual galaxies relative to the average for the entire cluster. This is a pretty tedious task, and at the time was quite difficult. As it turned out, the Coma Cluster is speeding away from us at an average velocity of 7,000 km/sec. But this is an average, and is like saying a swarm of bees is travelling away from us at some speed. At any given instant some bees will be travelling away faster than others owing to their motion within the swarm. In the same way, the motions of galaxies within the Coma Cluster means that some are travelling away from us at greater velocities than others, even though the entire cluster is moving away. If a galaxy within the Coma Cluster is moving away from us at 5,000 km/sec, the implication is that it is actually moving towards us relative to the cluster at 2,000 km/sec. Similarly, if a galaxy is observed to be moving away from us at 9,000 km/sec, this implies that it is moving away from the centre of the cluster at a speed of 2,000 km/sec. Neither of these galaxies, however, will escape from the cluster any more than the stars Oort observed will escape from the Milky Way. The cluster's gravitational field is more than strong enough to reign in such potential escapees. Given time, the gravitational field of the cluster is strong enough to turn the orbits of the galaxies around and drag them back into the centre. The important point is that the greater the spread in velocities – which indicates the total kinetic energy of the system – the greater the gravitational energy needed to keep the galaxies in tow. Since gravity implies matter, a comparison of the kinetic energy and the gravitational potential energy directly indicates the mass of the system.

VIRIAL THEOREM

As mentioned earlier, despite the tremendous difference in scale between an individual galaxy and a cluster of galaxies, the relationship between kinetic and gravitational potential energy is a common element in the search for dark matter at Galactic and galaxy cluster scales. The difference is that when it comes to more irregular systems such as clusters of galaxies, in which the motions are less defined, the calculation is a little more difficult. Nonetheless, Zwicky was able to calculate the gravitational mass of the Coma Cluster by using a technique known as the virial theorem (which follows from the Jeans equations used by Oort). This relates the amount of a system's kinetic energy due to the motion of the component objects within it – that is, the galaxies in a cluster – to the gravitational potential energy needed to prevent it from flying apart.

MASS–LUMINOSITY RELATIONSHIP

Armed with a knowledge of the gravitational mass of the Coma Cluster, Zwicky estimated the amount of mass accounted for by the individual galaxies. Now he may not have been quite right, because at that time astronomers were unsure about the mass of an individual galaxy in terms of stars (and this remains a problem even today). However, just as Oort and others assumed that the luminous mass of the Galaxy could be determined by looking at the total

Figure 3.3. The Virgo Cluster of galaxies contains more than 2,000 galaxies and dominates the Local Supercluster. This picture shows the western central region of the cluster. To the lower left is the giant elliptical galaxy M87, also known as Virgo A, believed to be the dominant member and close to the dynamical centre of the cluster. The two bright elliptical galaxies at right (west) are M86 (left-most) and M84. (Figure 3.4 shows a deeper view of this part of the cluster obtained with the Mayall 4-metre telescope at Kitt Peak National Observatory.) This picture was created from sixteen images taken in June 1995 and January 1997, using BVR filters, at the Burrell Schmidt telescope of Case Western Reserve University's Warner and Swasey Observatory located on Kitt Peak, near Tucson, Arizona. These images are of different parts of the cluster and are mosaiced together, which is why the picture is an odd shape inside the frame. The most obvious effect of this mosaicing is a change in sky noise along the boundaries where different numbers of original exposures contributed. The total image size is 122×86 arcminutes (approximately $2 \times 1°.5$) – an area twelve times the size of the full Moon. (Courtesy NOAO/AURA/NSF.)

Figure 3.4. The Virgo Cluster of galaxies (see Figure 3.3) includes the elliptical galaxies M84 and M86. Thousands of galaxies form a rich, loose, irregular cluster which is not strongly concentrated towards the centre. This image was obtained with the Mayall 4-metre telescope at Kitt Peak National Observatory in 1974. (Courtesy NOAO/AURA/NSF.)

number of stars and using the stellar mass–luminosity relationship, from his photographs Zwicky measured the brightness of individual galaxies and produced a fairly realistic estimate of the total luminosity of the cluster. He then assumed that the stellar content of each of the galaxies was more or less the same – which astronomers still have to do today.

RESULTS OF STUDYING THE COMA AND VIRGO CLUSTERS

All in all, it was a reasonably elaborate treatment: measure the brightness, velocity and location on the sky of individual galaxies within a cluster, and then deduce the luminous mass for the cluster. When he did this, Zwicky found a discrepancy between the luminous mass and the gravitational mass. If the cluster contained luminous mass alone, the velocities of the individual galaxies should have sent them flying to the far reaches of the Universe aeons ago. This was obviously not the case, and so something was exerting enough gravity to hold the cluster together. But what was astonishing – so much so that it took decades for the astronomical community to fully accept Zwicky's result – was the amount of matter involved. Whereas Oort had surprised everyone by saying that the Galaxy contained perhaps twice as much matter as could be seen in the form of stars, Zwicky's results revealed a far higher discrepancy between luminous and gravitational mass. According to his calculations, the Coma Cluster contained as much as fifty times the luminous mass, in some unseen form. The Coma Cluster was saturated with dark matter.

Zwicky did all this in 1933, and three years later Sinclair Smith, at Mount Wilson Observatory, carried out a similar experiment with the Virgo Cluster. This is a much closer cluster of galaxies, and is irregular in shape. So massive and close is the Virgo Cluster that our Local Group of galaxies is being pulled towards it. (The Local Group of galaxies consists of two large spirals – the Milky Way and the Andromeda galaxy – along with a number of dwarf elliptical and irregular galaxies.) Like countless other galaxies, both past and future, our Galaxy may one day succumb to this 'Virgocentric flow' and join its ranks. While the central region of the Virgo Cluster is dominated by a giant elliptical galaxy, the bulk of the cluster's visible mass is in the form of a hot (10–100-million-degree) gas that permeates the cluster. But Smith revealed the Virgo Cluster's dark secret. When he carried out the same calculations on the Virgo Cluster that Zwicky had performed on the Coma Cluster, he concluded that it contained an astonishing one hundred times more dark matter than luminous, and that this material probably lies between the galaxies.

The virial theorem was a very direct argument for dark matter. Since Zwicky's work on the Coma Cluster, this same technique has been applied to many other clusters, yielding similar results. Far from shedding light on the nature of the Universe, astronomers were finding it to be an increasingly dark and mysterious place. In fact, astronomers are still using arguments not that much more sophisticated than Zwicky's original.

CONTRAST BETWEEN OORT AND ZWICKY

Until the last 25 years or so, the dark matter argument presented by Oort seemed to have had more impact on most astronomers than that presented by Zwicky, even though the problem presented by Zwicky was much more extreme. Some people have speculated that this was due to the difference in the personalities of the two astronomers. The contrast between these two great men could not have been sharper. There was Oort – the most pre-eminent Dutch (and probably world-wide) astronomer of his time, much beloved and revered by his fellow astronomers. Then there was Zwicky – an irascible fellow who liked to disagree with the deeply held scientific beliefs of his fellow astronomers, no matter what those beliefs might be. He was, in short, both stimulating and irritating. But perhaps another explanation is that Zwicky's results – which required as much as fifty times the dark matter as luminous matter – strained the imagination of astronomers. Even though all astronomers for the last fifty years have been aware of Zwicky's results, they were more comfortable with ignoring them than with accepting the conclusions they required.

By the end of the 1930s, two major figures in the astronomical community had proclaimed that much of the Universe consists of dark matter. And yet the subject of dark matter lay dormant for four decades, while all around major astronomical discoveries and issues involving luminous matter achieved prominence. These included the formation and subsequent rivalry between the Steady State and the Big Bang theories of the origin of the Universe, and the discovery of pulsars, radio galaxies and the brightest objects in the Universe, the quasars.

Quasars are the active cores of very distant galaxies, so distant that they are among the oldest objects in the Universe. They are also immensely bright, emitting more energy than a hundred supergiant galaxies. Black holes and the microwave background radiation (the echo of the Big Bang) were predicted and discovered. Important as they are, all these discoveries skirted around the dominant form of matter in the Universe.

Then, in 1970 a discovery reignited interest in dark matter, fuelling it so much that by the end of the 1980s astronomers around the world were carrying out detailed analyses of dark matter in individual galaxies. Whereas Oort had suggested that dark matter lurks in the disks of galaxies, Ken Freeman and others used observations from radio telescopes to measure the way galaxies rotate, and discovered the first evidence that dark matter exists in giant halos that completely surround galaxies, including the Milky Way.

4

Dark halos

HOW TO MEASURE DARK MATTER HALOS

Oort, Lindblad and Stromberg discovered the rotation of the Milky Way by studying the motions of stars. Oort's later observations of the motions of stars perpendicular to the Galaxy's disk allowed him to measure the mass of the Milky Way, and led him to what we now believe to be the erroneous conclusion that the disk of our Galaxy contains dark matter. But what about other galaxies? How massive are they, and do they contain dark matter?

Other galaxies are so far away that their congregations of stars appear as nebulous patches of light to all but the largest telescopes, and so trying to answer questions about their mass by observing their individual stars is out of the question. We need an alternative. If it is a spiral galaxy then it is pretty straightforward: since all the stars are rotating more or less as a single mass, it is possible to measure how fast the galaxy is rotating and to calculate how much gravitational mass is needed to hold it together. This was first attempted by Horace Babcock in 1939 for the galaxy M31, the famous Andromeda galaxy. He began by measuring the Doppler shifts at various places around the edge of the galaxy, and then compared these with the galaxy's average movement through space to reveal the rotational velocity. But of course, being able to convert the observed velocity of the edges of a galaxy into kinetic energy requires knowledge of the physical size of the galaxy, and this depends on the distance and the apparent size. Furthermore, to be able to determine the amount of luminous matter it is necessary to know the brightness of the galaxy, which also depends on its distance. When Babcock made his observations, the distance to M31 was poorly known. In the end he underestimated its distance and hence its size, which in turn led him to the conclusion that the ratio of luminous mass to gravitating mass was 50. Although this is now known to be wrong, he was at least heading in the right direction.

BEYOND THE VISIBLE DISK: THE 21-CM LINE

Babcock and his contemporaries were restricted to observing only the visible disks of galaxies, but it was not long before the dark matter problem was extended beyond this limit. All of the observations up to this time were made using the shifting of spectral lines in the optical part of the spectrum, simply because the ability to observe the Universe at other wavelengths did not yet exist. But as we said in Chapter 2, one of the major contributions Oort made was the development of Dutch radio astronomy; and you will recall that it was one of Oort's students, Henk van de Hulst, who predicted the existence of the 21-cm line of neutral hydrogen (discovered in 1951). This line is no different from the spectral lines that Oort and Zwicky used, except that it is found in the radio rather than in the visible part of the electromagnetic spectrum, and it comes from neutral hydrogen gas between the stars, rather than from stars themselves. The radio and visible spectra are part of a continuum called the electromagnetic spectrum, and are all subject to the same laws of physics. Spectral lines in the radio spectrum, therefore, experience the same Doppler shifts caused by movement along the line of sight. The fact that the 21-cm spectral line is produced not by stars but by the hydrogen gas within and around a galaxy is a real advantage. It enables radio astronomers to measure the Doppler shift, and hence motions, of different parts of a galaxy, regardless of whether or not there are stars there, as long as hydrogen gas is present in its neutral or atomic form.

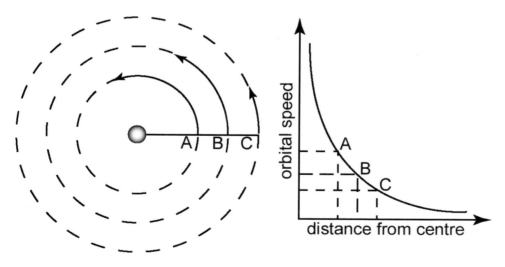

Figure 4.1. The principle behind rotation curves. Objects further from the centre of a rotating system – whether it be a solar system or a galaxy – do not have to rotate as fast in order to remain within the system. As a result, a plot of their orbital speed against distance from the centre of the system tends to flatten out at a predictable rate.

This kind of work is carried out with instruments such as the Australia Telescope Compact Array. A hydrogen map of the galaxy produces all the Doppler shifts of the radio waves emitted from different places in the galaxy, and these shifts can then be used to measure very precise velocities all over the face of a galaxy that contains hydrogen. (Most of the galaxies containing observable hydrogen are either spirals or irregular galaxies.)

If there was nothing more to galaxies than meets the eye, then the stars and, more importantly, the gas in the outermost regions of a galaxy, should be travelling slower than the stars closer to the centre: further from the centre of the galaxy the gravitational pull is weaker, and so matter need not travel so fast to resist the inward pull. A plot of the rotational velocity of the galaxy against distance from the centre produces what is known as a 'rotation curve'. In the absence of dark matter a galaxy's rotation curve should rise to a peak and then begin to drop, indicating that the rotation velocity of the galaxy becomes lower the further you move out towards the edge of the galaxy. This is the differential rotation that Oort discovered for our own Galaxy in the 1930s. What everyone assumed was that the rotation curves of galaxies held no surprises. In fact, during the 1960s many rotation curves were measured using optical techniques, by observing the ionised gas that is found in the inner parts of spirals. However, these rotation curves do not reach out to the outer parts of the galaxies, and indeed no real surprises were found until the 21-cm data began to appear.

THE FIRST SIGNS OF TROUBLE

While studying something entirely different, Freeman discovered that galaxies were not rotating the way everyone had assumed. Spiral galaxies have a very characteristic distribution of light, and the amount of light drops off exponentially from the centre to the edge (although to this day no-one knows why). Freeman determined analytically what the rotation curve of such a galaxy would be, and compared it with the rotation data that was then available. Rotation curves of galaxies can be measured only as far as the hydrogen gas extends (though this is usually much further than the visible edge). By the time the end of the rotation data is reached, however, it is possible to measure the mass within the last measurable radius.

In 1970, observations of the rotation of other galaxies were thin on the ground, but Freeman found enough to get him interested. He noticed that the rotation curves for three or four galaxies were simply the wrong shape based on the assumption that the galaxies were made of stars, gas, and nothing else. In fact, they were a miserable fit to the expected curves, as the observed rotation curves were not as expected. This meant that not only are the velocities of stars and gas not falling with distance from the centre, but are in some cases the matter is moving faster! Freeman published a paper pointing out that something was definitely wrong. The obvious explanation which he suggested was that these galaxies contained considerably more invisible matter than visible matter –

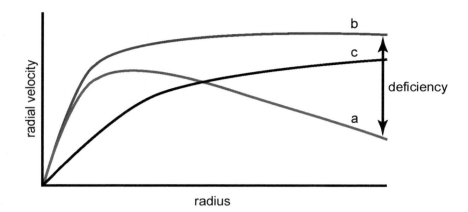

Figure 4.2. A plot of the expected and observed rotation curves for the Milky Way: a) the expected rotation from stars and gas alone; b) the observed rotation of the Milky Way; c) additions required from the dark halo to produce the observed rotation curve. Stars further from the centre of the Galaxy are moving faster than expected, resulting in a flatter rotation curve. The only explanation for this is a halo of dark matter surrounding the Galaxy and 'pulling' the stars around faster than they would otherwise orbit. At the distance of the Sun from the centre of the Galaxy, the stars are rotating at an average of 220 kilometres per second.

but not in the disks. It had to be further out beyond the stars and gas, in a halo of dark matter that completely surrounded the luminous galaxies. New 21-cm observations of the Andromeda galaxy around the same time, by groups in the UK and US, supported the view that something was wrong.

HOW TO SUPPRESS BAR STRUCTURES

Not long after Freeman made his announcement, another idea also spurred astronomers' thinking about the subject. It was a theory which, like Oort's work, is now believed to be not so relevant to the dark matter problem, but at the time it prompted astronomers to think seriously about dark matter. We mentioned in Chapter 3 that Hubble classified galaxies according to their shapes. Among the spirals, it was believed at that time that about one third show strong central bars – the barred spirals.

In 1973 two astronomers at Princeton University – Jeremy Ostriker and James Peebles – published a paper attempting to explain why so few spiral galaxies develop bars. In other words, what was preventing the majority of galaxies from developing bars? At that time, people had begun computer simulations of spiral galaxies, and had noticed that their models very often grew central bars. Stars like to hang around each other, attracted by their mutual gravitational pull, and their gregarious nature is contagious. Even a small clump of stars will lead to still

Figure 4.3. Visible light image of the barred spiral galaxy NGC 1300. Of the 75% of galaxies in the Universe that are spirals, about two thirds are barred spirals – so-called because of the stable bar structure across the centre. The lack of a bar in many spirals was once thought to be the result of a halo of dark matter, but it is now known that it is due to the random motion of stars in the centre of the galaxy. (Courtesy Pat Knezek, WIYN Consortium Inc., NASA, ESA and The Hubble Heritage Team (STScI/AURA).)

more clumping, which attracts yet more stars, and so on. The clumps grow due to gravitational instability, and it is this instability which drives the galaxy towards a bar shape, simply because it is a more stable structure than an evenly distributed disk of stars.

At the time, astronomers did not realise just how many spiral galaxies actually do have bars (about two thirds), so they were asking why the remaining spirals do not have bars. Believing that the majority of spirals did not have bars, Ostriker and Peebles argued that one way of preventing a spiral galaxy from developing a bar is to surround it by a very massive halo of dark matter. This would reduce the influence of the gravity of the stars in affecting the shape of the galaxy. The stars would, in fact, be like test particles that trace out the shape determined by the dark halo.

It is now realised that there are other ways to suppress bar formation. One way is to increase the random motions of stars in the inner parts of the disk, so that instead of having all the stars in the disk moving neatly around in circles, the random motions of stars can be increased towards the centre. This is what really happens, and during the 1980s Freeman and his colleagues at Mount Stromlo Observatory and the University of Groningen were able to measure the random motions of stars in the disk of a few spirals, including our Galaxy. Since then other astronomers have carried out similar measurements of other spirals. The result is that the velocity dispersion – the amount of randomness – is quite high towards the centres of galaxies, and it is this that can stabilise them against forming a bar shape.

We now know that in most galaxies the dark halo is very important for the structure and behaviour of the outer part of the galaxies; the inner parts of galaxies are dominated by the gravitational influence of stars. So what Ostriker and Peebles concluded – that galaxies are surrounded by dark matter halos – was correct, but perhaps not for the right reason. Although arriving at the truth in this way may not seem very orderly or planned, it happens all the time in science!

The 1970s were exciting times for the development of dark matter science. Following on from their ideas about dark halos, Ostriker and Peeebles, with their colleague Amos Yahil, put together the dynamical evidence that was available at the time regarding the masses of galaxies. It led them to the view that giant spirals like our own Galaxy really do have masses around 10^{12} solar masses – much higher than was generally thought at the time. Their evidence included neutral hydrogen rotation curves of three nearby spiral galaxies (M31, M81 and M101), measured by Morton Roberts and Arnold Rots, the motions of pairs of galaxies around each other, and virial mass estimates for small groups of galaxies. Their paper was published in 1974; but even though their data were fairly basic, it was a real landmark in the development of our ideas about dark halos around galaxies. Their mass estimates are close to the best that we can achieve today, thirty years later. As so often happens in science, other people were thinking along the same lines. In Estonia, Einasto, with his colleagues Kaasik and Saar, had also concluded from similar arguments, and in the same year, that spiral galaxies must be very massive.

THE 21-CM LIMIT

During the 1970s, Vera Rubin and her colleagues at the Carnegie Institution of Washington were accumulating high-quality optical rotation curves for spirals, which usually showed flat rotation curves. (Recall that we expect the rotation curve to fall with increasing radius, if the gravitational field of the galaxy comes from the stars alone.) Although flat optical rotation curves can rarely provide conclusive evidence for dark matter (because they do not reach out far enough from the centre of the galaxy), these data did keep people thinking about the problem. But the real confirmation came in 1978 with the publication of a PhD thesis by Albert Bosma at the Kapteyn Laboratory in Groningen. Using the Westerbork Radio Synthesis Telescope he compiled 21-cm rotation curves for about twenty spirals, and showed that almost all of these rotation curves were flat out to the edge of the 21-cm data.

So, by the end of the 1970s it was pretty clear that there was a big problem, the essence of which was that either the whole Newtonian formulation of gravity is wrong (an idea that is taken seriously by some, as we will see later), or that galaxies are surrounded by giant halos of dark matter. But the problem may be

Figure 4.4. Westerbork Radio Synthesis Telescope. (Courtesy Netherlands Foundation for Research in Astronomy.)

yet bigger. The rotation curves derived from 21-cm lines simply indicate the rotation velocity of the galaxy as far as the hydrogen gas extends. It is extremely probable that there is even more mass that cannot yet be detected, because it lies beyond the edge of the detectable hydrogen. So the questions remain. How far can the dark matter halo be probed? And is there any way of probing beyond the hydrogen disk of a galaxy?

BEYOND THE 21-CM LIMIT

When examining the hydrogen distribution in spirals, it seems to have a pretty sharp edge. The density of the hydrogen drops away steadily over scales measured in many kiloparsecs, and then it drops dead over the last kiloparsec or so, and vanishes. The reason this happens is not that the hydrogen is not there, but simply that we cannot see it. So, why would the hydrogen suddenly become invisible? Probably because it is being ionised by ultraviolet radiation. When an ultraviolet photon strikes a hydrogen atom it separates the positive nucleus (a single proton) from the negative electron. The origin of this ultraviolet radiation is a puzzle, but there are two distinct possibilities. One is that there is quite a lot of ultraviolet radiation produced by quasars in deep space; but there are fairly strong limits on just how much of it can be produced in this way, and at the moment it seems that this is not the right answer. Another possibility is that most spirals, in their inner regions, have a fair amount of star formation going on, which produces a large amount of ultraviolet radiation.

It might be wondered how the hydrogen in the outer regions of a galaxy can be influenced by the star-forming regions in the centre of the galaxy. After all, galaxies are very flat, and hydrogen in the outer regions cannot look straight back at the star-forming regions in the centre of the galaxies, simply because there is a lot of dust closer in which would absorb the ultraviolet radiation. Despite the seemingly flat nature of galaxies, the outer regions of a galaxy are

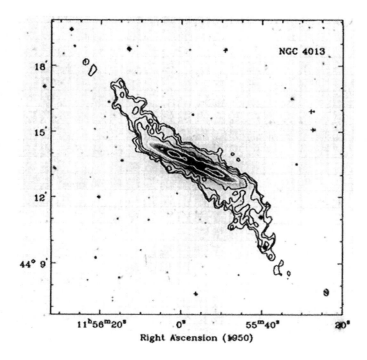

Figure 4.5. NGC 4013 is a nearby edge-on spiral galaxy. The dark greyscale image near the centre shows the optical light (as a negative, so that stars and bright regions appear darker). The optical light comes mainly from stars in this galaxy. The contours show the distribution of neutral hydrogen gas in this galaxy, superimposed on the optical image. The hydrogen gas layer warps away from the plane of the optical light, especially in the outer regions of the galaxy. This galaxy is one of the most warped systems known. The hydrogen observations were made with the Westerbork Radio Synthesis Telescope in the Netherlands, by R. Bottema and colleagues.

usually not all that flat, as the hydrogen disk aften warps up or down a little so that the disk is shaped like the brim of a slouch hat. As Joss Bland-Hawthorn of the Anglo-Australian Observatory pointed out, this offers the hydrogen at the edges of the galaxy a more direct look at the inner parts of the galaxy where all the star formation is taking place. This warping has been known for a quarter of a century, and although warping of the star-plane is rare, warping of the hydrogen layer is common, and it is now known that most galaxies are warped.

While there are plenty of ideas as to what causes these warps, nobody is yet convinced as to why it happens. It could be, for example, that the halo of dark matter is slightly flattened and therefore has a preferred plane of symmetry. If this plane is not parallel to the plane of the stars in the galaxy, then it will tend to pull the hydrogen in the outer regions of the galaxy towards itself – above the plane of stars where the dark matter is higher, and below where it is lower.

Figure 4.6. Visible-light image of NGC 4013. (Courtesy C. Howk (JHU), B. Savage (U. Wisconsin), N.A. Sharp (NOAO)/WIYN/NOAO/NSF.)

However, it remains one of the classic dynamical puzzles in astronomy, and despite some of the greatest minds having attacked it, it remains unresolved.

Whatever the mechanism, the idea that the hydrogen is in line of sight with the star-forming regions in the centre of the galaxy works out quantitatively rather better than some of the other explanations, although we should point out that this is all fairly new, and opinions one way or the other will change many times.

You may wonder why only the outer regions of the galaxy experience this ionisation. Why not the rest of the hydrogen in the disk of the galaxy that we can see? In fact, the entire galaxy suffers a dose of ultraviolet radiation from the centre, but those ultraviolet photons can only do so much ionising before they are used up. If a region is dense with hydrogen, the ionising photons are used up in the outer skin of the hydrogen layer, and the inner hydrogen, protected from the ultraviolet photons, stays neutral. . . and visible. But in the outer parts of the galaxy, where the density of the hydrogen is much lower, a point is reached where the number of ultraviolet photons is enough to ionise virtually all of the hydrogen and keep it that way. When that happens, there is a sudden drop in the amount of neutral hydrogen that can be seen in the 21-cm line.

This may be the reason why the edge appears so sharp, but it also means that beyond the neutral hydrogen edge seen with radio telescopes there is a region of ionised hydrogen. Now, an ionised hydrogen atom does not stay ionised all the time, but goes through a process called 'recombination'. When an ion – a hydrogen nucleus, or proton – and an electron meet, they join together and emit recombination radiation in the process. This produces a neutral hydrogen atom, which is very quickly ionised by another ultraviolet photon. This recombination radiation is visible as a group of spectral lines which can be observed. One of these lines (which is quite famous among amateur and professional astronomers) is the H-alpha line – the strongest hydrogen line which we can readily observe without going into space. The amount of H-alpha radiation observable from the outermost regions of galaxies is very faint, and until recently it was too faint to detect. Improvements in spectrographs, and an instrument called a Fabry–Perot interferometer, now enables astronomers to measure these faint levels of ionised hydrogen emission.

With Joss Bland-Hawthorn and other colleagues, Freeman has observed it with the 3.9 metre Anglo-Australian Telescope at Siding Spring Observatory in New South Wales. They have been able to push the rotation curves beyond the limits previously imposed by the neutral hydrogen. The first galaxy on which they used this technique was the big, nearby spiral NGC 253, and the technique has since been applied to M33, a famous spiral galaxy near the Andromeda galaxy. It is not yet clear how far this method can be pushed, as much depends on how far the hydrogen actually extends. There is a long way to go before the entire dark matter halo is encompassed by direct observation.

We can also extend our observations beyond the atomic hydrogen limit by using satellite galaxies as tracers of the gravitational field. Again, the random motions of these satellites around their parent galaxy prevents them from falling into the parent, so we can use their motions to measure the gravitational field of the galaxy out to a few hundred kiloparsecs from the centre of the parent. Dennis Zaritsky, of Steward Observatory, pioneered this technique. It now seems fairly clear that the dark halos of spirals (including our own) extend out typically to 150 kiloparsecs or more.

DARK MATTER IN ELLIPTICAL GALAXIES

So much for spirals; but what about the giant elliptical galaxies? Spirals rotate, and by studying the rotation curves astronomers have deduced the existence of dark matter halos. But the motions of stars in elliptical galaxies are not so straightforward. How do you measure the mass of such passive – that is, non-rotating – galaxies? There are several techniques for estimating the masses of ellipticals, and although none of them are quite as direct as those applied to spirals, two techniques stand out.

One of them is to look at the motions of objects very far out in the halos of these ellipticals: either planetary nebulae or globular clusters, whose velocities can be measured one by one. Planetary nebulae play such an important role in the search for dark matter that it is worth digressing to take a closer look at these beautiful objects.

IMPORTANCE OF PLANETARY NEBULAE

We have already seen how massive stars end their brief but brilliant existences in spectacular explosions called supernovae. But such stars are in the minority; most stars end their lives as peacefully as they began. Billions of years after being conceived within gigantic clouds of gas and dust, stars like the Sun die peacefully, surrounded by ghostly clouds that glow silently in space. For a cosmologically brief moment, these beautiful clouds in the sky are teaching astronomers about the future lives of Sun-like stars.

The story of our understanding of star death begins in the eighteenth century as astronomers were ploughing their way across the heavens using a new generation of large telescopes. Invented in the early sixteenth century, until the mid-seventeenth century telescopes had small lenses that revealed only the nearest, brightest objects in the night sky. Foremost among these celestial explorers was William Herschel. Herschel was an accomplished musician, but his true passion was astronomy. Using a home-made telescope he scanned the heavens in search of new, undiscovered jewels. In 1781 he found Uranus, the first planet to be discovered since antiquity, whose erratic wanderings eventually led to the discovery of the planet Neptune and one of the first examples of finding unseen, dark matter based on the behaviour of visible matter.

As they scanned the heavens, astronomers often stumbled across small round gas-clouds which looked much like planets. Seen through eighteenth-century telescopes, the appearance of these strange objects could fool an astronomer into thinking they had discovered a new planet or comet. The only way to confirm the discovery of a planet was to keep a close eye on the suspect, because planets orbit the Sun and appear to move against the background stars from night to night. The nebulae, on the other hand, remain fixed against the stars. Because of the superficial resemblance of some of these nebulae to planets, Herschel called them planetary nebulae. He was fascinated by them, and when he found one with a faint star at the centre he was convinced that the nebulae were clouds of gas surrounding a star. This was a tremendous insight considering what was known of stars at the time, which was virtually nothing. Two hundred years later, astronomers have confirmed Herschel's picture: a planetary nebula is a cloud of gas surrounding a star... a star at the end of its life.

All stars – the Sun included – derive their energy from nuclear burning, initially of hydrogen into helium. The nuclear reaction produces tremendous amounts of energy, some of which we receive on Earth as light and heat. The outward rush of the energy generated deep within the core of the star is enough to counteract the crushing weight of the star, keeping it inflated for billions of years. With a finite supply of hydrogen fuel, however, a star's lifetime is limited. For a star like the Sun, about 5 billion years after its birth its supply of hydrogen begins to dwindle. With less energy being pumped out, the core of the star contracts under its own gravity. The increase in density at the core increases the energy output of the star, causing the outer layers to expand and cool. Swelling to several tens of times larger than the Sun's diameter, the star becomes a red giant, outshining the Sun a hundred times. The increase in density at the core is enough to ignite the fusion of helium into oxygen and carbon, while the conversion of hydrogen into helium continues in a shell around the core. Now the star resembles an egg: a yolk of helium surrounded by a white of hydrogen. Eventually even the helium is exhausted, but the star is not heavy enough to burn the oxygen and carbon to form yet heavier elements. With no outward-rushing energy to keep the star inflated, its core collapses under its own weight. Again, the outer layers expand and cool, but this time the star enlarges several hundred times its original diameter.

Deep within the swollen star is the dying core – half the mass of the Sun, and now a thousand times brighter. Streaming away from the core is a tenuous stream of matter travelling up to 20 km/sec. The expanding gas cloud – once the outer layers of the star – slips the gravitational bonds of the core and drifts silently into space. As the gas expands it cools, growing cold and dark and hiding the star within it. After thousands of years, however, the core collapses still further, heating to some 25,000 K. The prodigious radiation streaming outward from the star causes the surrounding gas to glow in much the same way that the gas in fluorescent tubes glows when electricity is passed through it. At its peak, the central star can reach 200,000 K – as bright as a thousand Suns. The glowing gas, with its hot central star, has become a planetary nebula, and it may last 100,000 years. Finally, the gas disperses and the star proceeds to its death as a white dwarf. Spectacular as planetary nebulae are, they are a mere wisp in space. A typical planetary nebula may have as many as 100,000 atoms in the same volume of space as a sugar cube. For comparison, the air you are breathing now has some 10,000,000,000,000,000,000 atoms in the same volume. So tenuous is the cloud that anything from the distribution of mass within the star to the density of the surrounding interstellar medium can cause it to adopt bizarre shapes. While most planetary nebulae are symmetrical, only about a tenth of them look really spherical.

Planetary nebulae do not live long, and their gaseous corpses dissipate within a few tens of thousands of years. Nonetheless, they have a unique characteristic that makes them visible while all else is dark. The physics occurring in the gas around the hot central star means that about 20% of the light of the star is emitted at a single wavelength, visible as a bright emission line – specifically that of doubly ignised oxygen [OIII] at a wavelength of 5007 Ångstroms in the blue-green part of the visible spectrum. This means that at this one wavelength, these little stars are incredibly brighter than anything else around them. Planetary nebulae have become marker buoys for the matter in nearby elliptical galaxies.

An alternative to planetary nebulae are globular clusters, which are so bright that they can be observed in nearby elliptical galaxies, although again large telescopes are required. There is a fair amount of work that is now being carried out on these, and the indications are that elliptical galaxies are also surrounded by huge amounts of dark matter. Of course, these techniques involve an assumption about how the stars are moving, and we do not really know that. With the spirals we know that everything is going around in circles, which makes life simple, but with the ellipticals we really do not know what those motions are.

A second, independent technique makes use of the fact that many elliptical galaxies emit X-rays from hot gas in their outer regions. This gas is not rotating, but rather is just a disorderly mess. Astronomers think that this is due to the stars in the outer regions of the galaxies, which throw off gas as they age. Due to their velocity, the gas is thrown into the interstellar medium, resulting in clouds of gas zooming along in different directions and colliding with gas from other stars moving in opposite directions. The energy that was once contained in the motion of the gas is converted to heat – so much heat, moreover, that X-rays are produced. Since elliptical galaxies are often quite bright in X-rays, the distribution of this hot gas can be studied, and therefore what kind of gravitational field is needed to retain it.

Many of the elliptical galaxies studied in this way are the big, central galaxies in the centres of clusters. For example, M87 in the Virgo Cluster and NGC 1399 in the Fornax Cluster are supergiant ellipticals and the easiest ones to measure because they have lots of X-ray emissions, globular clusters and planetary nebulae. While neither method – looking at the distribution of hot gas and studying the motions of globular clusters and planetary nebulae – is totally conclusive, they all yield the same result.

SHAPE OF THE DARK MATTER HALO

The shape of dark matter halos remains a mystery. They could be flattened like a half-inflated beach ball, or they could be spherical. The only thing that is known for sure is how much mass exists inside some radius. It is an interesting issue, because when astronomers carry out simulations of the early Universe coming together and they build halos dynamically, these halos always appear a little flat. In fact, the typical halo is rather prolate; that is, it is shaped very slightly like a rugby ball. So far, however, it has been very difficult to confirm the theory observationally. Freeman has been involved in such attempts, and can testify that it is a disheartening process!

One of the earliest attempts to measure the shape of the dark matter halo involved studies of polar ring galaxies. These unusual galaxies consist of a more or less normal disk galaxy encircled by a ring of stars and gas that rotates in a plane perpendicular to the disk. A total of 157 polar ring galaxies are known, and although just how they are formed remains a mystery, it is possible that they are the result of the merger of two normal galaxies. The disk of the main galaxy is more or less familiar, although almost completely depleted of the gas from which stars form. Star formation goes on in the perpendicular ring, which sometimes contains a few billion solar masses of neutral hydrogen.

In 1983 Vera Rubin, François Schweizer and Brad Whitmore published an exciting paper in which they examined the velocities of material in the polar ring of the galaxy AO136-0801. Just as in the case of the disk of the galaxy, the rotation curve of the polar ring was flat, and the conclusion was that the ring was surrounded by dark matter that extended beyond the visible edge of the galaxy, at least three times further! Another spectacular example of this phenomenon is the galaxy (or rather, pair of galaxies) NGC 4650A. This object consists of two galaxies ploughing into one another so that now they look like two giant, perpendicular wheels with a common centre.

Astronomers had high hopes that having two concentric galaxies with disks in opposite planes would reveal the shape of the dark matter halo surrounding them. Freeman was one of them. The idea was to measure the rotation rates of the galaxies in the two perpendicular planes, which would reveal the forces in those dimensions and hence the amount of dark matter in those directions. Despite initial hopes, the results are ambiguous, and even with more recent and much improved data, it is still difficult to interpret properly. The reason is that there are different ways of reproducing what you see. For example, as Magda Arnaboldi and Françoise Combes have shown, some models produce a flat dark matter halo, but the halo can lie in either one plane or the other, and it is very difficult to determine which explanation is correct.

FLARING OF THE HYDROGEN DISK

Another trick astronomers have tried in their attempt to determine the shape of dark matter halos is to look at the flaring of the hydrogen in the disks of galaxies. The hydrogen layer in a galaxy is not absolutely flat like a pancake. It is flat in the inner regions, but further out it starts to flare out – that is, becomes thicker – probably because the amount of mass, and hence gravitational pull, in the disk is becoming weaker. As a result, the layer of hydrogen becomes thicker in the outer regions of the galaxy. The degree of flattening depends very much on the shape of the dark matter halo. If the halo is very flat, it will pull the hydrogen down towards the disk of the galaxy more than if the halo is spherical. Astronomers have tried to argue from the amount of flaring at different distances from the centre of the galaxy. At first this technique looked promising, but it too has offered more than one solution.

Astronomers are therefore over a barrel as far as the shape of the dark matter halo is concerned. It would be satisfying to be able to say whether the cosmological simulations are correct or not. It also becomes quite a practical issue with one of the most ambitious searches for dark matter, Project MACHO (whose story we will tell later), since the interpretation of the results of this project depends on the shape of the dark matter halo. Until then, we will have to be satisfied with the fact that dark matter halos not only exist, but dominate the behaviour of galaxies.

5

We are surrounded!

We have seen how astronomers have been able to measure the dark matter halos of other galaxies, and it seems pretty clear that dark matter exists in all galaxies. Ours is no exception. Despite the early indications of Oort, the Milky Way's dark matter does not lie in the disk – although, as we said, it would please more than one astronomer if it did – but in a halo just like the other galaxies. How do we know?

ROTATION CURVE OF THE MILKY WAY

The Sun is about 8 kiloparsecs from the centre of the Galaxy, and out to about 20 kiloparsecs the same sort of rotation techniques used for other galaxies can be applied to yield very similar results. Out to this distance, the rotation curve is flat – a clear sign of a dark matter halo. A rough correlation between mass and distance from the centre of the Galaxy emerges: the mass is 10 billion solar masses multiplied by the radius in kiloparsecs. (Here we use the British billion – 1,000 million.) So by the time we reach a radius of 20 kiloparsecs we have about 200 billion solar masses within a diameter of 40 kiloparsecs – already about three times as much as we can account for from the starlight. In other words, we are already well on the way to a large amount of dark matter in the Milky Way.

ESCAPE VELOCITY ARGUMENT (HALO STARS)

Beyond about 20 kiloparsecs, however, we hit a snag. Because the galaxy is rotating, everything we look at outside the Sun's orbit is moving across our line of sight, and there is virtually no component at all along the line of sight. In this case, Doppler shift measurements of stellar motions are practically useless, so we have to find other methods of weighing the Milky Way. Rather than bemoaning the fact that we live inside the Galaxy, astronomers take advantage of it.

What we are really looking for are markers outside the 20-kiloparsec limit that will reveal the motion of matter (and hence the mass) around the Milky Way. So what is out there? The Galaxy's stellar halo. This, however, should not to be confused with the dark matter halo, but is rather the very faint luminous halo that contains some of the oldest stars in the Galaxy. But you do not have to look far out in the halo for halo stars, for a few come to us! Of all the stars near the Sun, most belong to the disk and are simply orbiting the Galaxy with us. But there is a minority of stars – about one in a thousand – that come from the halo.

The luminous halo is a more or less spherical aggregate of stars which is hardly rotating at all. While the disk is maintained by rotation, the luminous halo is held in shape merely by the random motions of the stars within it. The halo stars are fast-moving, energetic stars, buzzing around the Galaxy like a swarm of bees. Now, because of these random motions, at least some stars will purely by chance pass through the neighbourhood of the Sun. When they do so, they pass through quite quickly.

The reason why this is so useful for determining the gravitational mass of the Galaxy is this. If you want to know how much mass exists in the Milky Way,

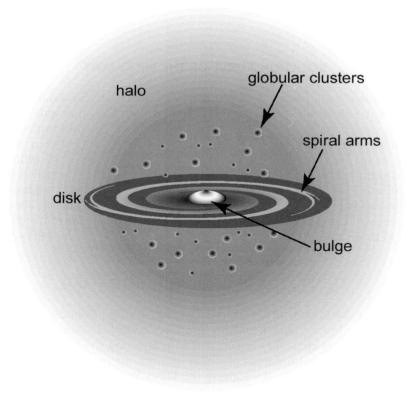

Figure 5.1. The main parts of the Milky Way. A typical spiral galaxy, the Milky Way is home to some 200 billion stars, most of which are in the spiral arms. At the centre of the Milky Way is a supermassive black hole with the equivalent of 3 million solar masses.

simply measure the velocity of the fastest star passing through the solar neighbourhood and from that calculate how much mass is needed to keep it bound to the system. This is called the 'escape velocity' argument, because it relies on the fact that if these stars had reached the Galaxy's escape velocity – the velocity they would need to escape the gravitational clutches of the Milky Way – they would almost certainly have already escaped. The fact that they are fast-moving and yet still nearby is a powerful argument that they are trapped by the gravity of the Milky Way.

What if the star is an interloper, not part of the Milky Way? True, this is a possibility. Stars do exist between the galaxies, as we shall see later in our search for the nature of dark matter. But we must accept that if a star is passing through the solar neighbourhood in the disk of the Galaxy, then the odds are very good that it belongs to the Galaxy.

The escape velocity only provides a lower limit to the Galaxy's mass, however. For one thing, although these stars are moving very fast, they may still be moving well below the true escape velocity from the Galaxy. Although stars with velocities up to about 650 km/sec are seen passing through the solar neighbourhood, to escape the clutches of the Milky Way a star may in fact have to reach 800–900 km/sec; but for some reason, such stars have either not yet been observed or simply do not exist. Another point is that there are probably stars in the Galaxy that are so energetic that they never visit the solar neighbourhood. So for both those reasons, the escape velocity argument can only provide a lower limit. Nonetheless, the lower limit is about 300 billion solar masses – a 50% increase on the mass of the Galaxy determined from the rotation curves.

OBJECTIVE PRISM REVEALS HALO STARS

But how on Earth can halo stars be found in the first place? And how can they be distinguished from the disk stars? Fortunately, there are a couple of ways. One is to look for nearby bright halo stars – which is fairly obvious, particularly when studying cooler stars. Being old, the halo stars formed long before the Galaxy was littered with the heavy elements that were later manufactured in supernova explosions, and as such they have specific chemical characteristics. In particular they are metal poor, which means they have heavy-element abundances – calcium, magnesium, iron, and so on – less than a tenth that of the Sun. So astronomers began taking spectra of a lot of stars and searched for those that have very weak spectral lines.

Now, this is not quite the 'needle in a haystack' task that it seems to be. Using a Schmidt telescope, astronomers were able to take the spectra of many stars at once. Schmidt telescopes are strange, often conical, instruments that have a sky-end aperture smaller than the main mirror, and yet their optical design provides them with the capability of producing sharp images over a wide field of view. They are therefore ideal for conducting all-sky surveys in relatively short periods of time. We have already seen how starlight can be broken into a spectrum to reveal its spectral lines, and this is usually achieved with a spectrograph. However, to find a rare kind of star, a more efficient approach is to use a glass prism that fits over the front of the telescope. These 'objective prisms' are enormous. For example, to use a Schmidt telescope such as the United Kingdom Schmidt Telescope at the Anglo-Australian Observatory, a 48-inch diameter prism would be placed over the front of the telescope. The objective prism splits the incoming starlight before it hits the film, and the resultant photograph shows the tiny spectra of many thousands of stars, each spectrum revealing the chemical composition of a star.

There are a number of techniques for distinguishing metal-rich and metal-poor stars in the resulting photograph. One is to simply look at the spectra, and many astronomers have done just that. Place a grid over the photograph so as not to miss any stars, and simply examine every single spectrum, one after another, looking for metal-poor stars. A more modern technique is to use computerised measuring machines to locate and analyse the spectra on the photographic plate. Either way, once a metal-poor suspect star has been found, a more detailed spectrum is taken with a larger telescope to confirm its nature. Frustratingly, halo stars are really quite rare, and many objects have been found that are not the elusive prey sought by astronomers. Thanks to automation, however, the task is now much easier, and while this modernisation was a long time coming, it was driven by astronomers' intense interest in metal-poor stars.

PROPER MOTIONS REVEAL HALO STARS

Another way of finding halo stars makes use of the fact that they move so quickly. As discussed in Chapter 1, the movement of a star across our line of sight is known as its proper motion. This can be measured by taking two photographs of the same stars as far apart in time as patience allows, or, more realistically, comparing a current photograph with one taken much earlier, usually by someone else. Comparing the relative positions of the stars reveals fast-moving stars: the faster the star is travelling in the plane of the sky, the further it will have moved between acquisition of the two photographs. Some of the halo stars have apparent motions on the sky of 0.5 arcsec per year, or even more. This may seem like a tiny amount – it is less than 3,500 times smaller than the diameter of the full Moon – but is actually not difficult to measure. It is quite a large movement in the scheme of things, and is the result of the fact that they move so fast through space. Such an apparently fast-moving star indicates one of two things: either it is nearby (the closer a star is, the faster its apparent movement across the sky), or it is physically moving very quickly through space. If it is the latter, it could be a halo star, worthy of closer inspection.

LOOKING FOR HALO STARS IN THE HALO

Impatient with waiting for halo stars to appear (observationally, not physically) in the solar neighbourhood, some astronomers decided to go to the source and began searching in the luminous halo. The technique is the same as when searching for nearby metal-poor stars, although one simple thing makes the search for halo stars in the halo difficult: they are a long way away, and are very faint. Nonetheless, astronomers have managed to discover them out to distances up to 100 kiloparsecs. By studying their random motions, the mass of the Galaxy, and hence the proportion of dark matter, can be determined. This is the measurement of stellar pressure again, but in the outer reaches of the Galaxy, and again provides some idea of the total mass. And it extends to about 500 billion solar masses out to an average radius of about 50 kiloparsecs! Now we are up to 2.5 times the mass determined from the rotation arguments.

TIMING ARGUMENT

Beyond the realm of our Galaxy lies another useful tool for helping us weigh the total mass of the Milky Way: the Andromeda galaxy. This magnificent galaxy – the largest in the Local Group – is a spiral not unlike our own. At a distance of about 750 kiloparsecs it is the closest large galaxy to the Milky Way. Since it is inclined at an angle of about 15° from edge-on, it looks like an oval-shaped nebulous patch of light in the northern constellation of Andromeda. We are indeed fortunate to have such a spectacle accessible to even the naked eye.

The Andromeda galaxy and the Milky Way are approaching each other at about 118 km/sec. Now, when considered, this is unusual, as the vast majority of galaxies are travelling away from us, drawn apart by the expansion of the Universe. So, what is drawing the Andromeda galaxy and the Milky Way together? The answer is gravity, which means matter, and lots of it. The two galaxies were probably born together, along with many other small galaxies that now make up the Local Group. While Andromeda and the Milky Way undoubtedly originally went their separate ways in the beginning, as part of the general expansion of the Universe, their combined mutual gravitational pull was sufficient to halt this separation and then reverse it! It can be immediately seen that there is now a method of determining the total mass between the two galaxies that is needed to achieve this titanic feat.

This technique was invented by an English astronomer, Franz Kahn, and a Dutch astronomer, Lodewijk Woltjer (another student of Oort), in 1959. It is called the 'timing argument', because it requires an assumption about how much time was needed for the two galaxies to reverse the Universal expansion and begin racing towards each other. In other words, an age for the Universe has to be assumed. If the Universe is old, then there has been plenty of time to turn the galaxies around and pull them towards each other. This implies that less mass was needed to reverse the recession. On the other hand, if the Universe is young, there has been less time to achieve the same feat, and that means that the galaxies needed to be more massive to pull themselves together in a much shorter time.

Knowing a precise age for the Universe was still some way off when Kahn and Woltjer were doing their work, but let us take an age of around 18 billion years as an example. This is rather on the long side – the currently accepted figure is 13.7 billion years – so it will yield a conservative estimate of the total mass of the two galaxies. Using this age, the total mass between the two galaxies turns out to be 4,000 billion solar masses. Our share of that is a little less than half – say 1,500 billion solar masses. This is nearly eight times the mass determined from rotation curves, and fifteen times the mass indicated by the Galaxy's luminous matter! Similar arguments can be applied for smaller galaxies that are also approaching us, to obtain surprisingly similar results. Needless to say, this is one of the things that makes astronomers feel more confident in their research into the amount of dark matter in the Milky Way system. A mass of 1,500 billion solar masses means that the flat rotation curve for the Galaxy should extend out to beyond 100 kiloparsecs. This is no real surprise, since the luminous stellar halo extends out to this distance, and so it is reasonable that the dark matter halo does likewise.

When all these arguments are put together, they produce a relationship between radius and mass mentioned earlier: for our Galaxy, the Galactic mass equals its radius in kiloparsecs times 10 billion solar masses, and this relationship holds out to at least 100 kiloparsecs. In that sense we know rather more about our Galaxy than we do about any other galaxy, simply because we have the advantage of living inside it, allowing us to apply these other arguments. With most galaxies, once we run out of hydrogen for measuring rotation curves, that is it. The estimates stop.

What all this means is that the dark mass of our Galaxy divided by the luminous mass is at least a factor of 20. But the study of ordinary spirals like our Galaxy does not produce numbers nearly as big as that, simply because when the hydrogen runs out not all of the mass has been measured. The hydrogen goes out beyond the stars, but the galaxies extend much further than that. When the rotation curves for typical spirals are measured, the dark matter:luminous matter ratio between 1:1 and 5:1, and only in a few extreme cases is it as large as in the Milky Way. The Milky Way is, therefore, a better representation of the real situation for a lot of galaxies. Smaller rotating galaxies yield similar results, and so the true dark matter:light matter ratio in galaxies is likely to be at least 20:1.

THE MAGELLANIC CLOUDS AND GALACTIC DARK MATTER

In recent years, astronomers have been able to use another nearby galaxy for determining the proportion of dark matter in the Milky Way... but only just. By observing the transverse motion of the Large Magellanic Cloud and comparing this with an existing knowledge of the motion along our line of sight, astronomers have been able to determine the Large Magellanic Cloud's true

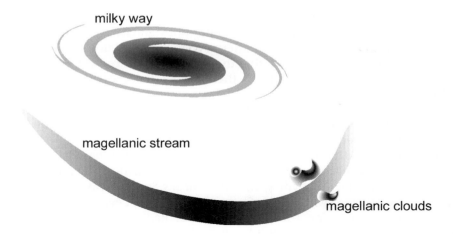

milky way

magellanic stream

magellanic clouds

Figure 5.2. Schematic diagram showing the structure of the Magellanic Stream. Although discovered in 1976, it was not until 1998 that clear evidence emerged for the tidal origin of the stream: the gravity of the Milky Way is stripping hydrogen gas from the two smaller galaxies.

space velocity. Now, that in itself does not reveal anything about the mass of the Milky Way. But another factor now enters the equation: the Large and Small Magellanic Clouds are themselves interacting.

In appearance, the Milky Way's two largest satellite galaxies, the Magellanic Clouds, seem to be peacefully orbiting their huge spiral companion. These big, bright, and beautiful galaxies grace the dark southern skies, complementing the brilliant arc of the Milky Way. But their peaceful appearance hides a stream of violence, as they are slowly being destroyed by being too close to the Milky Way. Evidence of their slow but inevitable demise is in the form of a trail of hydrogen debris marking their passage. This 'Magellanic Stream' extends for tens of thousands of light-years through space, forming a giant arc that extends over a 100° on the sky.

The Magellanic Stream, as it came to be called, was discovered in 1976 by Australian astronomer Don Mathewson, former Director of Mount Stromlo and Siding Spring Observatories. What Mathewson found was an enormous stream of neutral hydrogen gas trailing behind the two clouds like the smoke from a burning ship. The discovery of such a huge amount of gas 'leaking' from the two galaxies had a profound implication: the Magellanic Clouds were, in a sense, dying. Hydrogen gas is the 'food supply' of a galaxy, it is what stars are made from, and stars give a galaxy luminous life. But stars do not last forever and have to be replaced, so if a galaxy's store of gas is removed there is nothing else from which to make new stars. The Magellanic Stream represents gas being depleted from the Magellanic Clouds, so in a sense the galaxies are starving to death.

While the Magellanic Stream is clear evidence that the Magellanic Clouds are being destroyed, confusion existed over precisely how this is happening. One theory, based on tidal arguments, suggests that the Stream is the result of a tidal interaction in which the Milky Way's enormous gravity is wrenching material from the Magellanic Clouds. An alternative theory is known as ram-pressure stripping. It is based on the idea that the Stream is produced as the Clouds hurtle through the tenuous gas that fills the Galaxy's halo, sweeping gas from the Clouds like long hair in a breeze. This halo gas forms a hot corona that surrounds the Galaxy out to a distance of at least 70 kiloparsecs, and is directly observable by studying the X-ray spectra of quasars. Until recently it was difficult for most astronomers to decide which theory was correct.

In 1998 a breakthrough occurred that all but confirmed the tidal theory of the origin of the Magellanic Stream. The discovery was made during a survey called HIPASS – the HI (neutral hydrogen) Parkes All Sky Survey. This survey is a scan of the entire southern sky in search of large-scale structure beyond our Galaxy, and is being carried out using the 64-metre Parkes Radio Telescope and a unique instrument called the Parkes Multibeam Facility. Such a project is unprecedented, because of the massive amount of observing time needed. Surveying point by point across the sky with a single radio telescope and receiver would have taken a tremendous amount of time. The Parkes telescope can ordinarily see only a tiny part of the sky at any one time – about a quarter of the size of the full Moon – and so an all-sky survey like HIPASS would previously have taken thirty years.

Figure 5.3. The Magellanic Stream as seen in the neutral hydrogen 21-cm emission line. The brighter regions show denser hydrogen. The two large bright regions to the left of the figure are the Large Magellanic Cloud (upper) and Small Magellanic Cloud (lower), connected by a bridge of neutral hydrogen. The Magellanic Stream extends over about 150° on the sky. (Courtesy Mary Putman.)

The Parkes Multibeam Facility changed all that forever. What makes it special is its ability to 'see' large areas of the radio sky and study what lies there at several wavelengths simultaneously. Its wide-field capability is due to thirteen radio feeds at the focus of the giant dish rather than the single feed on other radio telescopes, and it can observe thirteen points spread over more than a 1-degree field of view. This allows astronomers to carry out large-scale radio surveys much faster than normal. Rather than taking thirty years, the Multibeam Facility achieved the job in three years!

The other feature of the Multibeam Facility is its ability to record the spectra of radio objects in the same way that astronomers have been working with visible light for decades. Astronomers can record the redshift of galaxies almost instantaneously, so that the recession velocity and hence distance to the galaxies being studied can also be quickly calculated. The product of these two innovations is a rapidly growing three-dimensional map of the galaxies, many of which have never been seen before. HIPASS is primarily a survey of gas in galaxies lying beyond the Local Group of galaxies, but as the programme got underway it was realised there was a huge amount of information about gas closer to home riding on the back of the main data. Of particular interest were data about fast-moving clouds of hydrogen dubbed 'high velocity clouds', or HVCs. The origin of HVCs was a mystery, and so the HIPASS team put together a PhD project to investigate the phenomenon. This coincided with the arrival of a new PhD student at Mount Stromlo Observatory, Mary Putman, now Assistant Professor at the University of Michigan. Her thesis programme was to better understand the distribution, origin and environment of the HVCs.

While Putman was carrying out this work she made a serendipitous discovery. The region around the Magellanic Clouds is one of the most interesting regions to explore in neutral hydrogen, and she reduced and combined the data about the Magellanic System as soon as it was available. The result was an abundance of new information and an interesting feature running ahead of the Magellanic Clouds on the sky – the long sought-after leading arm of the Magellanic Stream predicted by the tidal theory.

The initial observations revealed that the leading arm is large – at least about 25° long, about a quarter of the length of the trailing stream discovered by Matthewson. It has a radio brightness about a twentieth of the trailing stream, and so if the trailing and leading stream are at the same distance (which is impossible to know for sure) then this brightness ratio is equivalent to a mass ratio. This does not present a problem for the tidal theory, howeve. If the tidal interaction was a pure two-body interaction, the leading and trailing arms would be roughly equivalent. The fact is, it is a three-body interaction (two Magellanic Clouds and the Milky Way) and that breaks the symmetry, just as theory predicts.

There are various tidal theories that explain the Magellanic Stream, and no doubt all of them will have to be modified to account for the leading arm. But our Galaxy is involved in all of them. By taking these theories and the space velocity now known for the Large Magellanic Cloud, it is possible to estimate the mass of the Galaxy out to the Large Magellanic Cloud, about 50 kiloparsecs. The result is about 200 billion solar masses – a little less than the general run of estimates, but still much higher than the luminous mass – and it depends very strongly on understanding how the Magellanic Stream is produced.

DARK MATTER IN THE LARGE MAGELLANIC CLOUD

The Large Magellanic Cloud also probably contains some dark matter, and recent rotation studies suggest that it, too, has a flat rotation curve. Such studies are complicated, as it is such a disorganised and messy system. Its rotation is pretty disturbed, and additionally it is sitting inside the dark halo of our own Galaxy. So it is probably carting its own dark matter halo around within our dark matter halo, which presents confusion and uncertainty. In short, it is not the ideal galaxy to study. There are plenty of galaxies of similar luminosity that are not compromised by such interactions that enable the study of dark matter in small, irregular galaxies.

The Milky Way and its environs consist of the spiral galaxy itself, globular clusters, the two Magellanic Clouds, and a swarm of satellite dwarf galaxies. Neat as this inventory seems, it is insufficient. If only a fraction of the Galaxy is in the form of luminous matter, we can no longer use luminous matter as a description of the Milky Way. But, as we will see in the next chapter, high as the dark matter content of the Milky Way is, it is nothing when compared with dwarf galaxies, whose stars are mere markers for the location of huge concentrations of dark matter.

6

Pieces of the Big Bang

ABOUT DWARF GALAXIES

Galaxies come in a huge variety of shapes and sizes. At one end of the galactic spectrum are giant ellipticals that lie almost exclusively in the middle of clusters of galaxies. These gargantuans contain anything up to 10^{12} solar masses and have total diameters up to 200 kiloparsecs. Galaxies like the Milky Way are far more common among the large galaxies. In contrast to the giant ellipticals, which account for perhaps a fifth of all galaxies, spiral galaxies represent three quarters of all the bright galaxies in the Universe. Spirals are rich in gas and dust, the ingredients of stars, while ellipticals are comprised mainly of old stars and have little, if anything, left from which to make new stars. The remaining 5% are the irregular galaxies with no discernible regular shape or structure.

These figures, of course, exclude a large number of galaxies that have not yet been mentioned. They are the dwarf galaxies, and when they are considered the tally is different. Some of these dwarfs (the dwarf irregulars) have gas and patchy star formation, like the spirals. Others are almost gas-free, and are called dwarf ellipticals. The number of dwarfs depends on the environment: lots in clusters of galaxies and not so many in loose groups. Our Local Group of galaxies is a typical loose group, and there the census is four spirals, no giant ellipticals, twenty dwarf irregular galaxies and about seventeen dwarf ellipticals. Some of the smallest and most commonly occurring dwarf galaxies are very faint and difficult to see, even on deep images. They contain at most a few million stars... and a lot of dark matter.

AARONSON'S PIONEERING WORK

The first hint that dwarf galaxies contain large amounts of dark matter was discovered by Mark Aaronson, who pioneered the field of dark matter halos in small galaxies. (Aaronson was a great astronomer who was killed in a terrible freak accident while observing with the 4-metre telescope at Kitt Peak Observatory in 1987.) He measured stellar velocities in a few of these small galaxies – which was very difficult to do at the time. He had measured only half a dozen stars when he noticed that the velocity spread – that is, the range of velocities of the stars, a rudimentary measure of the pressure that keeps galaxies and clusters inflated – among these stars was unexpectedly large. Astronomers did not take this result too seriously; after all, six stars do not produce a measure of stellar pressure. Also, some astronomers suspected that some of the stars might have been binary stars. When two stars are orbiting each other as in a binary system, an orbital velocity is involved. During a survey of the motions of stars within a galaxy, you might measure two stars that both belong in binaries. Binaries often consist of a bright star and a faint, sometimes invisible, companion. In these cases, you could take a spectrum of the binary but only see the light and lines of the brighter star, and so be measuring its velocity only. Now, two binary systems within a galaxy might have exactly the same velocity, but if the bright star in one binary happens to be approaching the observer and the bright star in the other binary is receding, it creates the illusion of large differences in velocity. Of course, over time the spectral lines of both binaries would be seen moving back and forth across the spectrum as each rotating pair of stars swings the brighter star towards you then away from you; but in a one-off survey of the stars' movements, there can be large apparent velocity differences that do not reflect the true motion of the binary systems. This dispute continued for years, but these days there are samples of more than a hundred of these stars in several of the nearby dwarf galaxies. Astronomers have measured them many times, and there have been no variations of the spectral lines that should be present if the stars Aaronson measured were binaries. Now astronomers believe that the stars in these little galaxies are really moving very fast, and this is a pretty direct indication of dark matter.

THE DENSITY OF DARK HALOS: KORMENDY AND FREEMAN'S WORK

More recently, Ken Freeman has collaborated with John Kormendy, of the University of Texas, to reveal the bizarre nature of dwarf galaxies: not only do they contain dark matter, but some of them have a hugely disproportionate amount, considering their tiny size. The two astronomers have both been involved in the study of dark matter in dwarf galaxies for some time. Several years ago they realised that the dark matter in these small galaxies seems to be extremely dense, and so they began trying to accurately determine the dark matter densities of halos surrounding galaxies of different luminosities. Various people had measured halo parameters in different ways, and they had to standardise all of them and clean up the data. This required minimisation of the sources of error, and in particular trying to evolve a consistent method of determining the distances to galaxies, which is always a painful procedure. When they had finished, however, some rather tight correlations emerged that had previously been masked by messy data. They discovered that big spirals have rather low densities of dark matter, but that smaller and smaller galaxies contain increasing densities of dark matter.

Like their larger cousins, dwarf galaxies come in a variety of shapes. Some of these smaller galaxies have more gas in them than others. If they have a lot of gas they are irregular and their star-forming regions give them a patchy appearance; if they do not have gas they are quite smooth systems and are usually called dwarf spheroidal galaxies. The dwarf irregular galaxies rotate, so it is fairly easy to measure their halo densities, just as we do for spiral galaxies. Others, like the dwarf spheroidal companions to the Milky Way, are supported mainly by their random star motions. But regardless of which type of dwarf the two astronomers looked at, the result was the same: small galaxies have a lot of dark matter, and the dark matter is very dense. In fact, it happens that the stars in some of these galaxies constitute only a miserable fraction of the total mass; and in the most extreme cases, the stars account for only a hundredth of the total mass.

OBSERVING DWARF GALAXIES

Observing dwarf galaxies is not easy. The starlight from some of these smaller galaxies is so faint – that is, the number and concentration of stars is so small – that they can barely be seen in photographs. But because they are so close, astronomers can measure the velocities of individual stars, using telescopes such as the 3.9-metre Anglo-Australian Telescope and the even larger 8-metre telescopes in Chile and Hawaii.

The determination of the velocity of individual stars requires measurement of their individual spectra. Since pointing a telescope at so many stars individually would be a time-consuming task, astronomers use multi-object spectrographs that measure the spectra of many stars at a time. For objects that cover a greater area on the sky, like nearby dwarf galaxies, a system of optical fibres is often used to channel the light of tens or even hundreds of stars into a spectrograph, which then disperses the starlight into a spectrum that can be analyzed. In the early versions of these fibre-optic systems, one end of each of the fibres was inserted in a hole in a brass plate that was fixed in the focal plane of the telescope. Each hole was positioned on the plate exactly where an individual star's images would be formed. Although the use of fibre optics enormously improves the efficiency of astronomical observations on the night – an important consideration as telescope time becomes increasingly expensive – a great deal of preparatory work is necessary to achieve a good result. In order to know where to place the fibre tips it is often necessary to measure a photograph of the galaxy using an instrument called a microdensitometer. This is a very painstaking business, during which many things can go wrong, and yet it is crucial that everything is done correctly. If the star's position is not measured accurately, the fibre will be mounted in the wrong place on the plate and will not 'see' anything.

Microdensitometers are rare precision instruments costing a few hundred thousand dollars each. They work by shining a bright light through the glass photographic plate containing the image of the galaxy to be studied. The plate is moved in two dimensions to an accuracy of a micrometre over this light, and

Figure 6.1. A dwarf spheroidal galaxy. Although tiny compared with their larger spiral and elliptical counterparts, these galaxies can contain hugely disproportionate amounts of dark matter. (Courtesy Mount Stromlo and Siding Spring Observatories.)

since the plate is like an ordinary negative, any changes in brightness of the light indicate a star's image. Once identified, the positions of the star images on the plate are then converted to Right Ascension and Declination coordinates. At one time there were two of these instruments in Australia: one at Mount Stromlo Observatory and the other at the Sydney Offices of the Anglo-Australian Observatory. Unfortunately, about thirteen years ago the one at Mount Stromlo was destroyed when lightning struck the building in which it was contained, while the Anglo-Australian Observatory's instrument simply died of old age. As a result, these days photographic plates have to be sent to England for measurement. To facilitate this, a deal was struck between Australia and England. In return for Australia operating the UK Schmidt Telescope at Siding Spring, the English astronomers would measure and archive all of the photographic plates.

In the early version of the fibre systems, the stars' positions were translated onto a 40-cm diameter, 3-mm thick circular brass plate into which tiny holes were drilled, each hole representing the expected position of a star with the plate mounted on the telescope. A few extra, larger holes were also drilled at the positions of brighter guide stars, for fibre bundles that allow astronomers to point the telescope accurately at the galaxy. Each of the smaller holes was reamed out using a jewellers' reamer, and fitted with a single optical fibre. Once the plate was secured in the telescope, the other ends of the fibres were attached to a spectrograph.

The use of fibre optics in astronomy has been pioneered at the Anglo-Australian Observatory, and its present incarnation is a massive survey instrument called 2dF – the 2-degree multi-fibre spectroscopic survey facility. The 2dF differs from the earlier instrument in that the fibres are positioned not by hand into a use-once-throw-away brass plate, but by a robot on a reusable plate. Furthermore, 2dF does this while the telescope is in operation. As one plate aims four hundred fibres at the telescope image, a second plate is being configured with another set of four hundred fibres. Each plate is mounted on the outside faces of a device called a 'tumbler', which resembles an hour glass. At the end of each observing session, the tumbler flips over so that as one plate is carried out of the way the second plate moves into the telescope's focal plane. As this second set of fibres begins work, the original is reconfigured by the robot positioner. All of this takes place at the top of the 3.9-metre telescope. Designed and constructed by AAO staff, one of 2dF's main tasks is to make three-dimensional maps of the structure of the Universe by mapping the redshifts, hence distances, to galaxies. So successful have AAO scientists and engineers been in their efforts that the European Southern Observatory commissioned them to build a fibre positioner for their 8-metre Very Large Telescope in Chile. This fibre positioner is now in operation.

But back to the dwarf galaxies. At night, when the telescope is pointed at the sky, the guide stars are centred and the astronomers begin their observations. Throughout the night the telescope is trained on the galaxy's stars. Rather than make one long exposure, however, astronomers usually make many short (half-hour) exposures. The reason for this piecemeal approach is that it allows them to clean up the data progressively. One problem with using CCDs is that they are subject to cosmic rays – high-energy particles that stream in from all directions in the sky – that contaminate the information. Such defects are much easier to correct by combining many short exposures. Although it involves a great deal of high-precision preparation, the system works extremely well.

By the end of the night, each fibre in the system will have allowed the creation of the spectrum of a single star. By studying the spectrum, astronomers can determine how fast the star is travelling compared with a reference star of known velocity. Repeating this process for all of the targeted stars creates a picture of the velocity distribution and hence the amount and distribution of mass.

DARK MATTER IN DWARF GALAXIES

Kormendy and Freeman compiled measurements of the amount of dark matter in forty-three galaxies ranging from the most luminous spirals to the faintest dwarf galaxies known. Their work confirmed and quantified the correlation between the brightness of a galaxy (in terms of stars) and the relative proportion of dark matter: the smaller the galaxy, the greater the proportion of dark matter within its visible boundaries. Our Milky Way is a large galaxy, and more than 90% of its mass is in dark matter; but in the smallest galaxies with just a faint scattering of stars, dark matter makes up an even larger fraction of the total mass. An exception to this rule is intermediate-luminosity ellipticals with around 20% of the luminosity of the Milky Way. Based on observations of planetary nebulae out to large distances from their centres, they seem to contain little if any dark matter. We do not yet know why the intermediate ellipticals seem to be so different.

Furthermore, while large galaxies like the Milky Way have far greater total amounts of dark matter than do the dwarf galaxies (simply because they are much more massive in total), the dark matter is far more concentrated in the dwarfs. We can barely see the faintest dwarfs – they contain hardly any stars – but the central dark matter density is about 1 solar mass per 30 cubic light-years – a hundred times greater than the dark matter density in a giant galaxy, and several times larger than the density of stars and gas in the disk of our Milky Way, near the Sun. These faint galaxies may look like gossamer, but in truth they are more like cannonballs.

These correlations provide new insight into how dark matter affects galaxy formation. In particular, they help us to understand the growing evidence that the smallest known galaxies, like the dwarf spheroidal companions of our Milky Way that have only 1/100,000 of its luminosity, have massive dark halos. Astronomers were uncertain whether such massive dark halos in these tiny galaxies are normal or just some peculiarity, or even a mistake in how we interpret the observations. The new observations show that it is quite normal for a dwarf galaxy to have such high dark matter density, no matter whether it is a dwarf spheroidal or a gas-rich dwarf irregular galaxy.

WHY DO DWARF GALAXIES HAVE SO MUCH DARK MATTER?

But why should such frail-looking galaxies contain so much dark matter? The answer lies in the fact that the faintest dwarf galaxies are almost pristine remnants of the earliest time of galaxy formation. When you look at a dwarf galaxy you are really looking at a piece of the Universe as it was a very, very long time ago – perhaps when it was only 1/3,000 of its present age and was itself very dense – and it was about this time that galaxies began to fragment out. We know from theoretical arguments that the smaller galaxies formed first. Most of the fragments came together to build bigger and bigger galaxies, but in the process a large amount of orbital energy was injected into the fledgling giant galaxy. This tended to spread everything out so that the density became lower as the galaxy grew. In the case of our Milky Way, the dark matter density is only about 1/1,000 of the density in one of the dwarf galaxies. Not all of the dwarf galaxies were absorbed into the giant galaxies, however, and the little galaxies that we see now are those that escaped. This explanation of the high densities of dwarf galaxies is that whatever dark matter turns out to be, it is almost certainly pristine material from the beginning of the Universe. This is one of the facets that makes the study of dwarf galaxies such an exciting field of research.

The dark matter correlations argue against one theory about how most dwarf galaxies form. Some astronomers have postulated that dwarfs are produced when large galaxies interact: tides pull long tails of gas and stars out of the parent galaxies as they pass by, and small lumps of gas and stars in these tails can be held together by gravity and so survive the collision to look like little galaxies afterward. Basically this makes sense, and it is probably the way in which some dwarf galaxies are formed; but is it the main way in which the dwarfs form? The vicinity of our Galaxy is a good place to look for this kind of tidal remnant because of the interaction between the Milky Way and the Large and Small Magellanic Clouds (discussed in the previous chapter). It should be easy to distinguish tidal dwarf galaxies from primordial (very old) dwarfs, however, because large galaxies have low-density dark halos, and tides would stretch what is already there. So, tidally-made dwarfs should have lower dark matter densities than their parent galaxies. As we have seen, the smallest dwarfs have much higher dark matter densities than either the Milky Way or the Magellanic Clouds, so they are real galaxies going back to the early days of the Universe, and not just tidal debris. This means that the number of dwarf galaxies is not related to the limited number of galaxy collisions that have happened since big galaxies formed.

The ratio of luminous matter:dark matter in dwarfs may have been further amplified by the fact that less massive galaxies have a weaker gravitational hold on their contents, so the first stars that die in supernova explosions in dwarfs eject more of the remaining gas. These explosions have little effect on the dark matter, so small galaxies have retained less gas with which to make stars, resulting in low numbers of stars despite their high dark matter densities.

IS THERE A LARGE POPULATION OF UNDISCOVERED DARK GALAXIES?

Astronomers have long known that small galaxies are much more numerous than large galaxies, and so it is probable that there are many more yet to be discovered. In fact, the microdensitometers used by astronomers to prepare for studies of dwarf galaxies have also been used to discover them. By measuring the positions of stars on existing photographic plates, British astronomers were able to create contours of star densities and began noticing weak concentrations of faint stars that later proved to be truly associated with each other... in other words, galaxies. The smallest dwarf galaxies have so few stars that they are difficult to see against the foreground of Milky Way stars, and even more difficult to discover. Moreover, the more recent dwarf galaxy discoveries have been possible only through the use of computers analysing the positions of stars. To the human eye there is simply no correlation between the stars at all.

There is now a hierarchy of these satellite galaxies. Some of them are visible to the eye on images of the sky, but at the other end of the scale are others that are virtually invisible. Photographs of the sky where such a galaxy is supposed to be reveal no recognisable grouping of stars until the images have been analysed with a computer, but we know that the galaxies are there because we can measure the velocities of the stars which shows that they are moving through space together. Extrapolation of the trend towards more dark matter for smaller and smaller galaxies results in the possibility that galaxies exist in which there are so few stars that the galaxy is almost all dark matter.

WHAT SHOULD WE LOOK FOR?

Such dark galaxies are likely to have sizes of a few hundred light-years and masses of less than 10 million solar masses. Not surprisingly, such objects are exceedingly hard to find. As described above, one way may be to use computerised searches to look for very low star densities. Another may be to look for faint traces of hydrogen gas. A promising source of such discoveries is the HIPASS survey described in the last chapter. One of the first tasks of the Multibeam system was to look for galaxies 'behind' the Milky Way. Such galaxies are hidden at visible wavelengths because their light is absorbed by the gas and dust in our own Galaxy. Low-frequency radio waves (like the 21-cm radiation), however, pass right through the Milky Way. Therefore, gas-rich hidden galaxies that are invisible to optical telescopes can be seen clearly by radio telescopes like Parkes. Coupled with the wide and multispectral view of the Multibeam Facility, this provides a fast and efficient way of mapping the distribution of gas-rich galaxies.

One of the first major results to come out of the Multibeam survey was the existence of hundreds of galaxies on the other side of the Milky Way that were previously invisible to optical telescopes. These galaxies have now been shown to lie along sheets and lines that join up with structures visible on either side of the Milky Way.

A result which surfaced after astronomers analysed a mere 2% of the data flowing in from the Multibeam survey was the detection of a large number of dwarf galaxies in the nearby Universe. Most of these dwarfs are faint uncatalogued dwarf irregulars which can be seen on optical images of the sky, once it is known where to look. This is a useful discovery. Since there are so many more dwarf galaxies than giant galaxies, the dwarfs may be a better guide to the distribution of matter in the Universe, including the ubiquitous dark matter.

LACK OF DARK MATTER IN GLOBULAR CLUSTERS STILL A MYSTERY

The existence of dark matter in dwarf galaxies raises an important and so far unanswered question: why do globular clusters not contain dark matter? The logical flow of the dark matter problem starts at the galaxy level, but for reasons astronomers do not understand we do not see dark matter at scales smaller than galaxies. This is a puzzle, since there is no obvious reason why globular clusters should not have dark matter. Large globulars are about the same mass as the smallest dwarf galaxies – 10 million solar masses – although about a tenth the diameter of a typical dwarf galaxy. They even swim through the Milky Way's dark matter halo. It seems that globular clusters are formed when gas is compressed to high densities when clouds of gas collide. The formation process therefore involves only ordinary matter, and not dark matter. This is especially puzzling, because most of the globular clusters in our Galaxy are among the oldest known objects in the Universe, and were forming at about the same time as the smallest of the dwarf galaxies, which we know consist almost entirely of dark matter.

We have now come full circle in our exploration of dark matter in the Universe. We began with Oort's apparently mistaken identification of dark matter in the disk of the Galaxy, have sailed through the vast dark matter reservoirs of galaxy clusters, and have passed by the dark matter-saturated dwarf galaxies that surround the Milky Way. But there is one last and certainly spectacular source of evidence of dark matter that needs to be explored before we can summarise the dark matter content of the Universe: gravitational lensing.

7

Cosmic mirages

Co-author: Warrick Couch

HOW GRAVITY DEFLECTS STARLIGHT

One of the basic ideas behind Einstein's theory of General Relativity is that space is a physical entity that can be distorted by massive objects. Far from being an empty void through which planets and stars hurtle, space is as real and tangible as the pages of this book. And just as you can bend and fold the paper on which these words are printed, so too the massive objects can distort the fabric of space. The Sun, for example, distorts the space around it, creating a sort of dimple – a funnel-shaped hole with sloping sides which become flatter further from the centre. Rather than the Earth being held to the Sun like a ball on a string, it rolls around the edges of the hole like a marble around a roulette wheel.

Now this was a bizarre thought – then as well as now – but when the British astronomer Arthur Eddington heard of General Relativity he immediately recognised its importance and began making plans to test its predictions. One of its predictions was that, just as matter followed lines of least resistance in spacetime, so did light. This was not such a new idea. Newton himself, with his classical theory, predicted that light would be deflected as it passes some large mass. Essentially, he was just treating light as particles with a mass, and it can be calculated how much the particles will be deflected. However, the physics behind Einstein's prediction of the bending of light by gravity is somewhat different from Newton's theory.

Einstein's idea is simple enough to explain. Imagine you are in a ship out in space well away from any gravitational influence. You are strapped against one wall of the ship (to stop you floating around) and holding a torch so that it shines on the opposite wall. When the ship is stationary, the light beam follows a straight line to the other side of the cabin, where it strikes the wall at the same height from the floor as the torch. But if you fire the rocket engines, the ship will start to accelerate. Now, although light travels very fast, its speed is not infinite. This means that between the time a photon leaves the torch, travels across the

Figure 7.1. Albert Einstein. (Courtesy Jewish National and University Library, Hebrew University of Jerusalem, Israel.)

ship and strikes the opposite wall, the ship has actually moved slightly. The photon will therefore strike the cabin wall closer to the floor, below the height at which it started. Seen from the side, the beam of light would appear to be curved downward like the stream of water from a hose.

But here is the important point. Although Einstein did not express it in quite these terms, he pointed out there is absolutely no difference between the accelerating ship in space and a stationary ship on the surface of a planet. Gravity and acceleration are indistinguishable. The amazing implication of this is that light will be bent by a gravitating object in just the same way as a torch beam in an accelerating lift. A nice idea, but how could anyone possibly prove such a theory?

Eddington realised that the deviation would be small, even in the presence of a large gravitational field, so he decided to use the most massive object at hand: the Sun. Instead of a torch, he decided to use the stars themselves. According to theory, the light from background stars along the line of sight with the Sun would be bent, shifting their apparent position like a celestial mirage. The amount of deviation would be tiny – only about 2 arcseconds near the edge of the Sun – about the size of the full stop at the end of this sentence seen from a distance of more than 100 metres. But how can stars be seen so close to the Sun? They are normally invisible in the daytime sky, but during a total solar eclipse

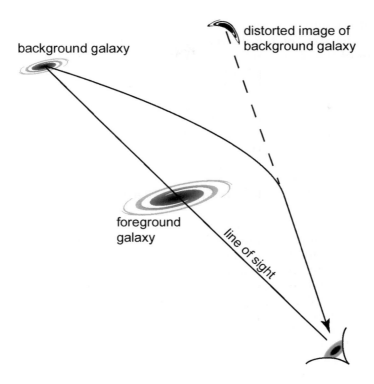

Figure 7.2. According to General Relativity, light from a distant source can be bent by the gravity of a foreground object. This was first observed by Arthur Eddington in 1919, but has since been seen in some spectacular examples such as the distortion of galaxy images by nearer galaxies.

the Moon obscures the Sun, and the stars become visible in the darkened sky. This would allow Eddington to photograph the stars near the Sun, and to then compare their apparent positions with their known fixed positions.

Eddington organised two expeditions to observe the total solar eclipse of 1919, himself leading one of the teams to the island of Principe, off the west coast of Africa. The other expedition was sent to northern Brazil. Two expeditions were mounted to improve the chances of seeing the eclipse. Should one team experience cloud during the eclipse, it was hoped that the other would have clear skies; but both teams experienced clear weather. When the eclipse photographs from the two teams were measured, the displacement of the star images near the Sun was almost exactly the same as predicted by Einstein.

Eddington reported his results to the public in November 1919. They were presented as a stunning confirmation of Einstein's theory. Headlines stated 'Einstein's Theory Triumphs' and 'Men of Science... Agog over Eclipse Observations'. The stars were deflected, and Einstein became famous. However, looking at the results of those observations today they appear to be rather marginal, as the sizes of the deflections were about equal to the errors in measurement of the photographic film. Since then, however, the observations have been repeated to greater precision, and Einstein's fame was indeed justified. The idea is now familiar: spacetime is curved by matter; light follows the curvature of spacetime. Ordinarily the curvature is slight, and so images of stars and planets in the night sky appear steady and unchanging. But when the alignment between a massive object and a distant light source is close enough, the deviation can indeed be seen.

THE MECHANICS OF GRAVITATIONAL LENSING

The ability of massive objects to alter the paths of light-rays provides an immensely important tool in the search for dark matter: gravitational lensing. If a distant luminous object – that is, an intervening gravitating object – and us are in a line and the right distances apart, then the image of the more distant object will be distorted and magnified. The way that this happens is similar to the way ordinary reading glasses converge light rays to a focus. Gravity does the same thing. If it can deviate single light-rays, it can also deviate bundles of rays, bringing them all to a focus. And if you have a focus, you have an image – a magnified, if distorted, image of the background object. Because the magnification is gravitationally induced, the intervening object that bent the light is called a gravitational lens.

Gravitational lensing is primarily a question of mass and geometry. In the simplest case of a light-ray passing some distance from a point mass on its way to the focus, the angle by which it is bent depends simply on the mass of the object and how close by the ray passes the object out in space. The deflection is usually not very large, and so just as a low-powered spectacle lens will have a long focal length, to form some sort of image at a given point using a gravitational lens very long distances are needed... cosmological distances. The optimum is when there are equal distances between us, the lens, and the background source. While this is unlikely to ever be the case, the fact is that if we were too close to the lens, or if the lens were too close to the background source, we would not see a lensing effect. However, as long as the distances are cosmological – that is, a billion light years or more between us and the background object – and the ratio between them not extreme, then we should see an observable effect.

THE EINSTEIN RADIUS

Normally, the alignment between a distant object and a gravitational lens is not perfect, and the best that can be seen is a distorted arc-shaped image of the object. But if we had a perfect situation where we had some distant object, a point mass and ourselves all in a straight line, the lensing effect would produce a perfect ring centred on the gravitational lens. The radius of the ring is called the Einstein radius. Even though most gravitational lenses are not perfectly aligned, the Einstein radius of individual arcs can nevertheless be measured. Why is this important? The Einstein radius is related to the mass of the lens, and so by measuring the Einstein radius from the shape and size of the arc, it is possible to measure the amount of mass in the lens contained within that radius. The implication for searches for dark matter are obvious, and we will get to them in a moment.

The Einstein radius is quite common in classic texts on gravitational lensing. The question is, when converted to seconds of arc on the sky, is it something that can be observed? In the case of the MACHO project (which will be discussed shortly), events involve stellar-mass gravitational lenses and background stars (point light sources) which result in Einstein radii measured in micro-arcseconds – hence the term 'microlensing'. The arcs themselves are, of course, too small to be seen. But that does not mean that light is not amplified. This is precisely what MACHO was set up to look for: more light rays arriving at the telescope than would normally be the case without lensing; in other words, brighter images. For clusters, the Einstein radius can be anything from a few arcseconds to 20 arcseconds, because of the large distances involved and the large masses in the clusters.

PROBABILITY OF LENSING EVENTS

As we have said, for a given light source–lens–observer geometry, the amount of lensing we see on the sky is simply a function of the mass of the lensing object (providing it is suitably compact): the bigger the object, the greater the amount of gravity-induced distortion. The exciting aspect for dark matter researchers is that since the amount of deviation depends on mass alone, the mass of a gravitational lens can be determined simply by measuring the amount of deviation. The beauty of this method is determination of the mass is not complicated by other factors that require assumptions and theories. There is no doubt about it. General Relativity provides exactly the amount of deflection expected for a given mass. Of course, the requirement that the geometry (source–lens and observer–lens distances) needs to be known should not be understated!

Any mass is capable of bending light, and therefore of behaving as a gravitational lens. It is simply a matter of whether or not the bending can be seen. This is where astronomy comes in. There are objects very nearby, such as the Sun, which cause a small, but observable deflection; while further afield, stars in our own Galaxy are capable of bending light from stars behind them, and so on. The critical factor in all these cases is that the background light source and the foreground object are both along, or very close to, our line of sight. When it comes to the stars of the Galaxy, we are essentially looking through a haze of stars at more distant background objects. Furthermore, the entire Galaxy is in motion, so that its individual stars are constantly moving across the sky

Figure 7.3. The Hubble Space Telescope peers through the centre of one of the most massive galaxy clusters known – Abell 1689. The gravity of the cluster's trillion stars – plus dark matter – acts as a 2-million-light-year wide 'lens' in space. This gravitational lens bends and magnifies the light of galaxies located far behind it. (Courtesy NASA, N. Benitez (JHU), T. Broadhurst (Racah Institute of Physics/The Hebrew University), H. Ford (JHU), M. Clampin (STScI), G. Hartig (STScI), G. Illingworth (UCO/Lick Observatory), the ACS Science Team and ESA.)

(admittedly very slowly as seen from Earth). The point is that with so many stars drifting across the field of view, the odds are rather good that at least a few of them will line up with some more distant object.

USING GRAVITATIONAL LENSING TO MEASURE DARK MATTER

When it comes to more distant examples – say, two clusters of galaxies lining up – the situation is different. In fact, as we described in Chapter 3, both Einstein and Eddington considered that the odds were so remote that while it was a nice idea, it had virtually no practical value (though in 1936 Einstein did publish the idea of gravitational lensing). Zwicky, on the other hand, reasoned that gravitational lensing would be visible if a galaxy were along the same line of sight as a more distant object such as a quasar. Because quasars are so bright and are generally seen at much greater distances than ordinary galaxies, they constitute bright background light-sources. As mentioned in Chapter 3, Zwicky extended this idea by proposing that not only was galactic gravitational lensing possible, but that the gravitational lensing of a distant quasar by a much closer galaxy or galaxy cluster would be an ideal way of studying the amount and distribution of dark matter in the foreground cluster.

As it turned out, Zwicky was right, though it was fully sixty years after Eddington's famous eclipse experiment that the first gravitational lens was detected. Moreover, gravitational lensing has turned out to be an ideal way of studying the masses of clusters of galaxies. Since galaxy clusters are among the most massive aggregates of matter seen in the Universe that are collected into relatively small regions of space – that is, they are small in comparison with the distances between us and background light-sources – they exactly meet the criteria for gravitational lensing mentioned earlier. As for background light-sources, the Universe is very rich in galaxies – particularly distant ones. If you can imagine a cone enveloping the cluster, with ourselves at the apex, beyond the cluster the cone envelopes a huge volume of space, increasing the chances of encountering a galaxy, or galaxies, which can be lensed. As we look deeper and deeper into space we see ever-increasing numbers of galaxies, providing a ready supply of background light-sources.

STRONG LENSING AND THE HUBBLE CONSTANT

When looking in the direction of rich galaxy clusters, astronomers often see what is referred to as 'strong lensing', which produces multiple images of a background object. One or more of those images is likely to be magnified and so will appear brighter than normal, though some of the multiple images may be fainter. Strong lensing offers a potential solution to an important problem in cosmology: the determination of the much-disputed Hubble constant (described in Chapter 2). Pinning down the Hubble constant has been a major challenge for astronomers for the last hundred years. Remember that Hubble showed that the further the distance of a galaxy, the faster its recession. This is because every galaxy is receding from every other galaxy. If we see two galaxies, one beyond the other, we will see the nearer galaxy moving away from us. The more distant galaxy will appear to be receding twice as fast as the nearer one because it is

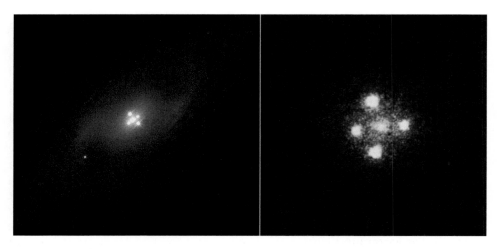

Figure 7.4. The Einstein Cross – the archetypal gravitational lens system, G2237+0305: (left) a wide-field view of the galaxy with the Einstein Cross at the centre; (right) a closer image obtained with the Hubble Space Telescope. (Courtesy NASA.)

moving away from the nearer galaxy, which is receding from us! Incidentally, it works both ways. The inhabitants of either of the other two galaxies will see precisely the same thing. All the galaxies in the Universe are moving away from each other as a natural consequence of the expansion of the Universe.

In 1929 Hubble showed that this is happening, and derived an initial figure of 200 km/s/Mps; that is, galaxies are receding from us at an extra 200 km/s for every megaparsec of additional distance from us. He underestimated the distances to the galaxies, however, and estimates of the Hubble constant fell sharply once it was known just how far away they really are. It is fitting that the telescope named in his honour – the Hubble Space Telescope – has pinned down the figure to a now widely accepted 72 km/s/Mpc. But does gravitational lensing offer support for this figure?

We mentioned earlier that one of the variables involved in gravitational lensing is the distance to the lensing cluster, which in turn depends on the speed with which the cluster is receding from us. The factor that links these two important values is the Hubble constant. Multiple images are formed from different light-rays taking different paths between the source and ourselves. Each of these paths has a different length, so when the background source is a quasar which is varying in brightness over periods of months to a year, the change in brightness of different lensed images provides a way of monitoring the relative distances, or path lengths between the quasar and the observer. This phase difference – the difference in arrival time of maximum or minimum brightness of the two images – is a function of the Hubble constant. In fact, it allows measurement of the Hubble constant, and so lensing is an independent way of measuring this all-important number. The critical factor in all of this is in understanding the lens very well, which is dependent on modelling it very well.

None of the systems studied so far have a simple point mass, which makes life a little more difficult. All of the results so far put the Hubble constant at a somewhat low figure: 50–60 km/s/Mpc. However, they also have very large uncertainties in them, arising from uncertainty about the nature of the lens. Nonetheless, it is an interesting way of approaching the problem, and in a world in which each method has its shortcomings, a new, independent method is very welcome.

WEAK LENSING

Much more common than strong lensing, though harder to detect, is 'weak lensing'. This does not produce multiple images, but rather simply distorts the image of the background light-source. All weak lensing does is change the shape of the background object, which is not obvious, and in fact can only be found statistically by measuring the shapes of many images in one direction in the sky and looking for distortion in a common direction. This distortion would be perpendicular to the line joining the image of the object and the centre of the mass causing the distortion. This effect has been observed through very careful analyses of images mainly obtained with the Hubble Space Telescope, though also using very high-quality ground-based images. Resolution is all-important in being able to detect these very small image distortions.

Strong lensing is much more conspicuous, and historically the first strong lens system to be found was the double quasar 0957+561. Astronomers had been looking for quasars which are conspicuously blue, particularly in the ultraviolet. A group led by Dennis Walsh obtained an optical image of the quasar 0957+561, and found that two star-like images were close by, separated by about 5.7 arcseconds. (For those familiar with the telescopic appearance of the double star α Centauri, it is about a third of the distance between the components). When they obtained a spectrum of one of these objects, its huge redshift revealed it to be the quasar. It is to their credit that they bothered to take a spectrum of the second object, as it was very probable that it was just a foreground star in our own Galaxy. In fact, the two objects have identical spectra and the same redshift. Two quasars so close together could still have been mere coincidence, as double quasars do exist. Confirmatory evidence came after long-term studies revealed that the two images varied in brightness in the same way, except that the variation of one image came in 420 days behind the other. The images had to be of a single quasar, and the quasar had to have been strongly lensed by a foreground object – probably a galaxy or a group of galaxies.

ABELL 2218

This is a special case, mainly because quasars being lensed in this way are quite rare, and are discovered at a rate of about one per year. Strong lensing of galaxy

Figure 7.5. The cluster Abell 2218 is arguably the most spectacular example of gravitational lensing ever observed. The images of galaxies beyond Abell 2218 have been distorted by the gravity of the foreground cluster. By analysing the degree and nature of the distortion, astronomers are able to map out the amount and distribution of dark matter in the foreground cluster. (Courtesy NASA, Andrew Fruchter and the ERO Team, Sylvia Baggett (STScI), Richard Hook (ST-ECF) and Zoltan Levay (STScI).)

clusters, on the other hand, is much more common, and one of the best examples is Abell 2218. This rich cluster of galaxies, lying at a distance of some 3 billion light-years, was made famous when it was imaged with the Hubble Space Telescope in 1994. The image is the result of a study carried out by a team of astronomers from four universities: Warrick Couch (University of New South Wales), Richard Ellis and Jean-Paul Kneib (Cambridge University), Ian Smail (Observatories of the Carnegie Institute of Washington), and Ray Sharples (University of Durham). It is now known as one of the best examples of strong lensing ever discovered.

The use of the Hubble Space Telescope to observe distant clusters was not motivated by gravitational lensing at all, but rather to study galaxy morphology (their shapes and structures). It was known that the galaxy cluster was undergoing strong evolution. The individual galaxies were much bluer, and so were in the process of star formation. The astronomers struggled to obtain the best images from the ground to determine whether there were any details in the structure of distant galaxies, but the galaxies are so far away and their images so small that details in their morphology remained frustratingly elusive. It was decided that the only way to obtain a clear picture of the morphology of the galaxies was to use a space-based telescope free from the distorting effects of the Earth's atmosphere: the Hubble Space Telescope. With a resolution ten times better than that possible on the ground, it would open up a whole new area of study in determining the morphology of distant galaxies.

It turned out that it is very common to see strong lensing through rich clusters of galaxies. One turned up in the very first cluster that the astronomers observed with Hubble Space Telescope. The image was an 'L' shape, and there were at least two images precisely in the middle of it. A symmetrical pair of images that were mirror reflections of each other was a tell-tale sign of a gravitational lens. It was soon clear that the astronomers could kill two birds with one stone: not only could they study distant cluster galaxy morphology, but they could also carry out a very nice study of gravitational lensing. Not only could the Hubble Space Telescope be used to discover gravitational lenses, but it could also allow measurement of their mass... and the study of dark matter.

The first images were obtained in 1992, before the Hubble Space Telescope had been repaired. Even with the telescope in a rather sick state, the potential for this type of work was remarkable, and it was obvious that after repairs it would have a wonderful future. The team had a jump on the rest of the astronomical community because it had time allocated to observe a number of high-redshift clusters. It was decided to optimise the future observations not just for studying galaxy morphology but also to make the most of the lensing side of it. Abell 2218 was well known as a strong gravitational lens from ground-based images. Its enormous gravitational field has distorted and magnified the images of the background galaxies – some three to four times further away – into several thin and faint arcs. It was an obvious object to study from space, and was thus made one of the targets.

The image of Abell 2218 was taken using the Wide Field Planetary Camera 2 in September 1994. When the image finally came in, the astronomers were blown away. They had no idea it would be so spectacular. It not only shows the foreground cluster of galaxies, but also a fantastic array of luminous arcs – the distorted images of distant galaxies – surrounding the galaxies of the foreground cluster. It has since become a bench mark – a text-book example of a gravitationally lensing cluster.

The team continued to study galaxy morphology, but following the Abell 2218 image, others who were expert at lensing analysis were brought in. In this case, lensing analysis is complicated by the fact that galaxy clusters are not simple point masses, but rather are extended mass distributions. Ellis and Smail had carried out some lensing work in the form of ground-based surveys, looking for any evidence of arcs that are the tell-tale signs of strong lensing, and they found a nice sample of clusters that showed evidence of lensing. But it was Jean-Paul Kneib, a French post-doctoral researcher, at the time at Cambridge, who was responsible for modelling Abell 2218 and carrying out the analysis of the lensed images of the cluster, Before joining the team Kneib had devoted all his research career to this sort of work, and so was already tooled up for the job. He had the programs on a computer to analyse the clusters, so was a very obvious person to become involved.

The subsequent analysis of Abell 2218 led to two results. Firstly, the determination of the mass distribution of the cluster, which involves a certain amount of trial and error. Initially, a model of what the lens might be like is created, based on properties such as the light distribution measured from the images. (The luminous matter is taken as a first indicator of how the underlying dark matter might be distributed.) Then a computer is used to simulate what would happen to different light-rays travelling from different background objects through the cluster – a process called ray tracing. The model is modified until it reproduces the image of the real thing. An enormous amount of information can be extracted from the Hubble Space Telescope images, enabling constraints to be placed on the model. Further modifications are introduced until the model produces a replica of the image. By this method it is possible to fairly well define the properties of the gravitational lens.

The Hubble Space Telescope image of Abell 2218 was a bonus for this type of work, since it shows a rare event: the appearance of multiple images of individual galaxies. An unprecedented seven multiple images were found. With strong lensing, each set of multiple images offered a chance to measure the mass of the galaxy cluster. This was possible because each individual image had a different 'impact parameter' – the distance between the line of sight passing through the lens and the background object, and the image. This distance varies depending on the mass of the intervening object. Since the light rays from each background galaxy take a different path through the foreground cluster, each one gives an independent measure of the mass of the foreground cluster. It is therefore possible to calculate the amount of mass that lies between the apparent location of the image and the true line of sight. What is important about multiple images is that they do not all lie at the same distance from the centre of the foreground cluster. With arclets visible at different places within the image, it is suddenly possible to measure the mass of the foreground cluster at different distances from the centre.

The second result from the image was the determination of the distance to the background galaxies. The lensing equations depend on the geometry of the whole affair: the distance between the source, the lens and the observer. The distance to the lensing cluster is easy to determine from redshift observations. What is not known is the distance from the lensing cluster to the background source. However, once the nature of the lens is determined from the computer simulations (described above) it is possible to calculate the distance to the background galaxies. Knowing the distance of the background galaxies from the lens, and thus from ourselves, it is possible to determine how long ago the light from the background galaxies left them – the look-back time. This allows an understanding of the properties of galaxies at that epoch.

Since this now famous image was obtained, astronomers have calculated the approximate distances of 120 arclets, allowing determination of the distances to galaxies fifty times fainter than those visible to ground-based telescopes. A gravitational lens can therefore be used as an enormous telescope to study objects that would otherwise be beyond the reach of even the largest ground-based telescopes. Without gravitational lenses, the distant galaxy clusters would be barely visible; but gravitational lenses such as Abell 2218 boost their visibility by factors of 10–50. The distant galaxies beyond Abell 2218 lie at a redshift of 1–

2, which means they existed when the Universe was less than a half of its present age.

The discovery of Abell 2218 added an extra dimension to the study of dark matter and opened up an entire cottage industry, and many rich clusters of galaxies have now been imaged with Hubble Space Telescope. Gravitational lensing offers yet another major source of evidence that most of these clusters are made of dark matter. From the gravitational lensing studies, it has been shown that 90% of the clusters are dark. There have been variations between clusters, but in each case the luminous matter constitutes less than 15% of the total cluster matter. The distribution of dark matter in the lensing clusters is similar to the individual galaxies, though there is some evidence that it is more strongly concentrated in the middle than the luminous matter.

Gravitational lensing is a wonderful way of studying dark matter. There is no complicated physics and there are no other variables to be accounted for, and so it is a very direct way of measuring the mass of the foreground cluster... all the mass, light or dark. Images of gravitational lensing also provide a direct record of the luminous material within the foreground cluster. By comparing the amount of luminous versus lensing mass, the mass–luminosity ratio for the cluster can be determined, and from that the amount of dark matter. Admittedly, clusters are not point masses, but rather, systems containing hundreds of galaxies sitting in a pool of dark matter, and so some assumptions have to be made about how that dark matter is distributed. But the theory is sufficiently well under control that very accurate measurements can be produced.

8

The baryon inventory

We have now scoured the Universe for dark matter, and along the way we have discovered evidence of its existence on a variety of scales – in dwarf galaxies, galaxies, clusters of galaxies, and so on. But because the methods used for studying the amount of matter at different scales are so diverse, there is always the possibility that one or all of the estimates could be wrong. A good way of checking these figures is to compare the amount of matter not at different distances in the Universe, but at different times throughout the history of the Universe. Let us say we compared the amount of matter in the Universe at three different epochs: the time of the Big Bang, when the Universe was about half its present age, and the present. If all three estimates agree, then we can be fairly comfortable that we know how much matter exists in the Universe, and where.

Ω (OMEGA) AS A COMMON UNIT OF MEASUREMENT

Before we can compare the amount of matter at different scales and times we need a common unit of measurement of the amount of matter in the Universe. Rather than trying to state it in kilograms, tonnes, or solar masses, cosmologists express it in a different way. The Universe, as we have said, is expanding as a result of the Big Bang. as a fraction of the total mass of the Universe, Ω, where $\Omega = 1$ equals the critical density. As we explained in the Prologue, the critical density is the amount of mass needed to halt the expansion of the Universe at some infinitely distant time; in other words, the expansion of the Universe will be forever slowing down but will never quite stop. With $\Omega < 1$ the Universe will expand unbridled forever, while $\Omega > 1$ means it will not only one day stop expanding, but will begin to contract again towards a Big Crunch. This last option is improbable, and while a critical-density Universe is predicted by an important version of the Big Bang theory called Inflation, by the late 1990s there were all sorts of arguments suggesting that the total Ω for matter is less than 1, and perhaps is as low as 0.2. The important point is that if the Universe contains anything less than the critical density, it will continue expanding forever. But here we are running the risk of getting ahead of ourselves in this story. The important thing here is that, despite the uncertainty of what fraction of Ω the

Universe actually contains, Ω is a convenient unit of measurement of how much matter – dark and otherwise – the Universe contains.

$$\Omega_b$$

Let us begin by looking at just how much of the Universe is in the form of baryonic matter; that is, how much of it consists of the familiar protons that make up our world and our bodies (see Appendix 1). Unlike dark matter, this also happens to be matter that we can see. The fraction of mass in the form of baryons we will call Ω_b.

BIG BANG NUCLEOSYNTHESIS

The Big Bang is the most widely accepted theory of the origin of the Universe. Although there are a number of versions, they all basically state that the Universe began as an immense fireball – a singularity of space, matter, time and energy that occurred somewhere between 10 billion and 20 billion years ago. (The currently accepted age of the Universe is 13.7 billion years.) During the first microsecond of the Big Bang, the Universe was a seething, blinding fog of matter and radiation. In that initial fireball, temperatures were so high that while particles had their individual identity, they readily changed from one type of particle into another. While we say matter is converted into energy inside stars – and this is a spectacular process – it nothing compared with the Big Bang. In that first microsecond, energetic photons slammed into each other to create matter: protons, neutrons and electrons, and their antiparticles. Almost instantly, however, these particle–antiparticle pairs would annihilate each other and produce pairs of photons. It was not long before the Universe expanded and cooled, sapping the photons of the energy needed to make protons and neutrons, although they still had enough energy to make electrons. All the while, protons and antiprotons, neutrons and antineutrons continued to annihilate each other, quickly reducing the numbers of each and producing radiation in the process. Within a second, photons lacked even the energy needed to make electron–positron pairs, and they too continued to annihilate each other. But not all the protons, neutrons and electrons were annihilated by antiparticles. There was a slight excess of particles over antiparticles – about 1 in a billion – that allowed our Universe to contain matter as well as energy.

About one hundred seconds after the Big Bang, baryonic matter was created. As the Universe expanded and cooled to a temperature of about 1 billion degrees, protons and neutrons seized power and took control of the Universe. The Universe then began behaving just like an enormous star, and it was still hot enough for nuclear reactions to take place: protons fused with electrons to form neutrons; neutrons and protons formed deuterium; deuterium nuclei combined to produce isotopes of helium. Other elements formed here too: lithium, beryllium and boron. Not too many of the heavy elements formed at this stage, however; they had to wait until the evolution of stars.

We shall return to the Big Bang in Chapters 13 and 14 to see how this baryonic matter ended up coalescing into galaxies and stars. What is important for astronomers is that computer models of this period in the history of the Universe predict relative abundances of these light elements that should exist today, and these abundances in turn depend on how much baryonic matter existed in the early Universe. Of particular interest to astronomers is the relative abundance of hydrogen (with a nucleus consisting of a single proton) to deuterium (a heavier form of hydrogen with a nucleus consisting of a proton and a neutron), which turns out to be a particularly sensitive measure of the amount of baryonic matter. The reason is that it is almost impossible to make deuterium in the modern Universe, as it can only be forged in the extreme conditions of the Big Bang. Furthermore, it is very easily destroyed, so that whatever is around today represents a lower limit on the amount that once existed. If astronomers can measure the amount of deuterium in the present Universe, they can calculate the upper limit on the amount of baryonic matter in the Universe.

OBSERVING BARYONIC MATTER

Making such observations is not an easy task, however. The trick is to look for absorption lines in the spectra of quasars. These lines are created by elements within clouds of gas that lie between us and quasars. Since quasars are so distant (you will recall that they are among the most distant objects in the Universe), their light passes through a great deal of the Universe. Astronomers can therefore measure the amount of deuterium in the Universe by looking at the relative strength of absorption lines due to hydrogen and deuterium. Now, the width of the hydrogen line is much greater than the line due to deuterium, and since they sit next to each other in the spectrum the hydrogen line frequently spills over the deuterium line, smothering it completely. To make matters worse, spectral lines are influenced by the motion of whatever produces them. For a single object there is only one movement of the spectral line, either towards or away from the observer. But with spectral lines produced by a cloud of gas, any motion within the cloud spells trouble. Gas moving towards the observer relative to the average motion of the cloud will cause the lines to be blueshifted, while gas moving away from the observer causes redshift. A turbulent gas cloud produces blueshift, redshift and 'no shift'! This results in a smeared spectral line, and so measurement is even more difficult.

Nonetheless, if the right quasar can be found in line with the right gas clouds, then there is a chance of measuring the relative hydrogen:deuterium abundances. More work is required, and many astronomers are making plans to study the problem further. Nonetheless, initial observations of the hydrogen:deuterium ratio have given a result for Ω_b, and while it is not an accurate number it is probably good to within a factor of 2 or 3. And what is the number? For the BBNS theory, Ω_b is about 0.04, or about 4% of the critical density. To put this into perspective, if the total mass of the Universe is only 0.2 of the critical density, then 0.2 of the mass of the Universe is baryonic (0.04/0.2 = 0.2). If the matter in the Universe really has the critical density, then the observations of deuterium together with the BBNS theory indicate that only about 4% is in the form of baryons.

OBSERVING Ω_b

That is the theory, but how much baryonic matter can we actually see out there, and does it match the prediction? What values of Ω_b are observed? We can measure the baryon abundance by adding up all the sources of baryons that we can see. A neat aspect of this direct measurement of the amount of baryonic matter in the Universe is that not only can we measure how much of it exists today, but also how much of it existed in the past. This is possible due to the finite velocity of light. Because light takes a certain amount of time to reach you from any given source, you are not seeing the object as it is now, but rather what it was like when the light left it. When you look at the Moon for instance, you are not seeing it as it is now, but as it was a little more than a second earlier when the light left its surface. Similarly, the sunlight you feel is actually almost ten minutes old, while the images of the nearest stars are years out of date.

Ω_b AT Z = 3

Admittedly, by the time we extend this concept to the Universe at large we run into trouble because of the unknown distance and age scale of the Universe. So rather than talk about the exact age of the Universe measured in years, astronomers prefer to talk in terms of redshift. As we have seen, the greater the distance to a galaxy, the more its spectral lines will have been redshifted due to the expansion of the Universe. Couple this idea with the fact that increased distance is equivalent to greater look-back time, and you can describe different epochs in the history of the Universe as different amounts of redshift. What we see nearby is at a redshift of zero, while the Coma Cluster, studied by Zwicky, is at a redshift of about 0.02. A redshift of 3 – that is, objects with spectral lines redshifted by 300% – is a good place to look at the past baryonic density of the Universe because it represents a look-back time of more than 80% of the age of the Universe.

   It is possible to estimate the baryon density at that time because most of the matter back then was in the form of huge filamentary clouds, like a froth of gas. The light from a distant quasar will have passed through this froth, which would have absorbed some of the light. The quasar's spectrum reveals that the wavelengths which were absorbed by the filaments show up as dark lines against the otherwise bright spectrum of the quasar. These dark lines come mostly from a hydrogen line called Lyman-α, and the spectrum now resembles a forest of tall trees, so the clouds are called Lyman forest clouds. What is the total baryon density at this period in the Universe's history? At this time, $\Omega_b = 0.04 \pm 0.01$, or about 4% of the critical density.

Ω_b IN THE PRESENT EPOCH

But what about the baryon density at the present epoch? At the present time – at a redshift of zero – baryons are found mostly in three forms: stars in galaxies, gas in clusters of galaxies, and gas in groups of galaxies (small collections of galaxies). A baryon census by Fukugita, Peebles and Hogan in 1998 showed that the amount of matter we have in the present Universe breaks down to stars in galaxies ($\Omega = 0.004$), gas in clusters ($\Omega = 0.003$), and gas in groups ($\Omega = 0.014$). This produces a total baryonic mass of about $\Omega_b = 0.021 \pm 0.008$.

But by far the most important result of recent times has come from a spacecraft called the Wilkinson Microwave Anisotropy Probe (WMAP). Launched in 2001, WMAP had the specific goal of establishing the parameters of the Big Bang: how old is it, what is it made of, and at what rate is it expanding? Looking from a vantage point 1.5 million kilometres from Earth, it produced an astonishingly detailed map of the cosmic microwave background – the afterglow of the Big Bang (which we will examine more closely in Chapter 13). In doing so, WMAP provided astronomers with the most accurate picture of how the Universe is put together. In short, it facilitated high-precision cosmology. Part of its phenomenally accurate list of cosmological data was Ω_b: 0.044 ± 0.04.

DOES Ω_b MATCH UP?

So now we have estimates of the baryonic density of the Universe from four different sources: BBNS predicts $\Omega_b = 0.04$; at $z = 3$, $\Omega_b = 0.04$; nearby, at $z = 0$, $\Omega_b = 0.021$; and now the WMAP result, $\Omega_b = 0.044$. Within the uncertainties, all three estimates are more or less the same, which means there are no major discrepancies, and so most of the baryons now and at $z = 3$ are all visible.

An important point here is that it need not have happened this way. It could have been, for example, that most of those baryons today were tucked away as white dwarfs, neutron stars, and so on, and we would never have seen them. But that does not seem to have happened, and it is a stroke of luck that we can see the baryons in the Universe now and at early times.

UNSEEN BARYONIC MATTER?

However, could there be still more baryonic matter out there that has not yet been seen? The baryon census in the present time ($z = 0$) found nothing in the voids between the frothy walls of galaxies at the largest scales. While there does not seem to be much in those voids in the way of galaxies, however, it could be that gas exists there that is invisible because it is at the wrong temperature to radiate at wavelengths that we can detect. Another source of missing baryonic matter is free floating stars in clusters or groups of galaxies. If stars had been ripped off galaxies as the galaxies tear past each other (cosmologically speaking), then they might also have been missed. Yet again, there could be baryonic matter out there in the form of massive compact halo objects (MACHOs). We will take a look at the MACHO project in the next chapter, but for now let us assume that these mysterious objects are made of baryonic matter and are familiar objects such as white dwarfs. If this is the case then a lot of baryons could also be tucked away there. But this is unlikely, for a very simple reason. If MACHOs are the galactic dark matter – and we know that the halo dark matter is about ten times the mass of the stars – then we finish up with MACHOs being ten times the total mass of stars – $\Omega_{MACHOs} = 0.04$ – on top of the visible stars and gas in the Milky Way. This would produce a total baryonic mass of $\Omega_{baryons} = 0.06$, which places the theory of BBNS in trouble, since it predicts an Ω of only 0.02. It is therefore improbable that MACHOs are baryonic.

BARYONIC MATTER IN GROUPS AND CLUSTERS OF GALAXIES

Let us take a closer look at the baryonic material in collections of galaxies. Galaxies tend to accumulate firstly into groups of anywhere from three to a hundred galaxies, while clusters can contain up to a few thousand galaxies. When groups and clusters are analysed using the same sorts of techniques employed by Zwicky, the result is an Ω indicated by the gravitating mass – stars, galaxies, dark matter, and anything that happens to be between the galaxies... the lot. We will call this gravitating mass Ω_{grav}. For groups of galaxies, Ω_{grav} is about 0.15 – about 15% of the critical density. Although clusters are more substantial objects, there are not as many clusters as there are groups and so they contribute only about $\Omega_{grav} = 0.03$ – 3% of the critical density. Groups of galaxies therefore really provide most of the gravitating mass that we know about in the Universe.

How much of this gravitating mass is baryonic? We know that the majority of the baryons in the present Universe is still in the form of gas, and yet this same gas contributes only about a tenth of the gravitating mass. When we look at individual galaxies like the Milky Way, however, the situation is rather different. We know that in individual galaxies most of the baryons are in the form of stars, but when this stellar mass is compared with the gravitating mass, again the answer is 0.1. In each case it seems that the total baryonic mass is about a tenth of the total gravitating mass. The rest is in some other form.

Figure 8.1. The vast majority of galaxies in the Universe gather together into collections ranging in size from a few galaxies, such as in the Hickson compact group shown here, to huge clusters of thousands of galaxies. However, there are far more groups (up to 100 galaxies) than clusters of galaxies. Groups of galaxies therefore make up the bulk of the gravitating mass in the Universe.

THE BARYON CATASTROPHE

Because at least a tenth of the mass in clusters of galaxies is visible, baryonic matter gives rise to what has become known as 'the baryon catastrophe'. In 1993 a group of astronomers led by Simon White (now director of the Max Planck Institute for Astrophysics, in Germany) published a paper describing the results of their study of the Coma Cluster of galaxies (the same used by Zwicky). Like Zwicky, they attempted to compile an inventory of mass in this cluster. They looked at stars and they looked at hot gas, and what they found was that the total mass of stars is about 1.4×10^{13} solar masses, while the total amount of gas amounts to about 13×10^{13} solar masses. That is, there is about ten times as much gas than stars. But the total mass found from looking at the movements of the galaxies – the gravitating mass – is about ten times the mass of the stars and gas – about 160×10^{13} solar masses. Of all the mass in the Coma Cluster, therefore, about 10% of the total gravitating mass is in the form of baryons, which is close to the 10% mentioned earlier. White's conclusions are supported by a number of other studies carried out by other astronomers on different clusters and superclusters of galaxies. For example, there are now at least twenty rich clusters of galaxies that have had their mass content and mass profiles measured using gravitational lensing methods. These studies show that around 80% of the matter in these clusters is dark, while a mere 20% is baryonic matter.

Here a conflict arises. The total mass of the clusters is determined by the motions of their galaxies, and this total gravitating mass can be compared with all the visible, hence baryonic, mass. In both Zwicky's and White's studies, the ratio is about 10:1 in favour of dark matter; that is, of all the mass in the clusters at least 10% is baryonic. However – and this is where the catastrophe occurs – Big Bang Nucleosynthesis as well as observations at a redshift of 3 and at the present time all indicate that the total baryonic mass is only about 4% of the total mass in the Universe, which was assumed to be the critical density. This suggests that there is a much larger fraction of baryonic mass in clusters (12%) than in the entire Universe (4%). At the time it was announced, the baryon catastrophe caused quite a lot of panic, but there are ways out of it.

One possible solution is that the baryons are more concentrated towards clusters than is dark matter, which is more evenly spread through the cosmos. However, theoretical simulations render this unlikely. It could be that the estimates of the gravitating mass are in error, but estimates based on other methods (X-ray and optical studies) yield very similar results. Perhaps something is wrong with the estimate of the amount of mass in the form of gas. After all, studies show that most of the baryonic mass in the Universe is in the form of gas. However, it is very hard to determine what could be wrong with these observations. Of course, it could simply be that the Big Bang Nucleosynthesis theory is incorrect, but it has been studied by dozens of cosmologists who have all reached more or less the same conclusion.

The easiest solution is that Ω for the mass of the Universe (Ω_{mass}) is not 1.0 at all, but is much lower. If we assume that the ratio between the baryonic mass and the gravitating mass in clusters (about 10%) is the same as the ratio between the baryonic density and the total mass density of the Universe, then the total mass of the Universe is 0.04/0.1 = 0.4 – about 40% of the critical density. This is not such a big shock, since there are other arguments that now point towards a similar conclusion. In fact, the baryon catastrophe is simply a consequence of believing that the mass in the Universe has the critical density. If this is not assumed then the problem goes away.

VIRGO AND COMA BARYONIC MATTER COMPARED

When the amounts and distribution of baryonic matter in the Virgo and Coma Clusters are compared, the values are similar. The total baryonic matter in the

Coma Cluster seems to be about 10% (1% in stars, 9% in gas), while the remaining 90% is dark matter. In the Virgo Cluster, the baryonic fraction is also about 10%, but there seems to be a higher proportion of stars (about 3%) compared with gas (7%). Nonetheless, in both cases – and they are very big cases, consisting of thousands of galaxies spread over millions of light-years – the results are appreciably the same: only about a tenth of the mass implied by the movements of the clusters is baryonic. Remember that stars only represent a relatively small repository of the Universe's baryonic matter.

There are two important points that arise in this part of our story. One is that the baryons are only a small fraction of the total mass in the Universe. It emphasises that the same fraction of baryons found at smaller scales of individual galaxies – on the order of 10% – is found at much larger scales of groups and clusters of galaxies. The second important point is that it seems we have a reasonable idea of the baryon density, as estimates from BBNS and observing the Universe at high and low redshifts all roughly agree. It is not as if 90% of the baryons disappeared from sight between a redshift of 3 and the present time, which could so easily have happened – but it has not happened.

And so here we sit in a Universe that is visually dominated by stars swirling among glowing gas. But the stars and galaxies and clusters of galaxies are only a small fraction of the mass of the Universe.

9

MACHO astronomy

Dark matter exists within the dark halos of galaxies; but what form does it take, and how can we find it? One possibility is the existence of concentrations of matter called massive compact halo objects (MACHOs). If halo dark matter exists in the form of MACHOs, it should be possible to detect their existence by watching for chance alignments between MACHOs and more distant, luminous objects such as stars. As a MACHO passes in front of a star it would gravitationally lens the star's image, causing it to brighten for a short time in a very characteristic way. Just how much and for how long depends on the mass of the foreground object, how far it is away from us and the star, and how fast it is moving. Therefore, the sudden and unpredictable brightening of a star image is a possible indication of something massive, such as a MACHO, passing between it and ourselves.

HISTORICAL BUILD-UP

The concept of microlensing goes back to Refsdahl, who carried out a huge amount of fundamental work on gravitational lensing long before it became such an active subject. As Refsdahl mentioned to Ken Freeman, he came up with the idea of gravitational microlensing in the 1960s, but did not follow it up. In the early 1980s a graduate student named Maria Petrou wrote a thesis on the subject while at Cambridge. Petrou was working on Galactic dynamics, and one of the things she pointed out was that if there is Galactic dark matter in the form of compact objects in the halo, then they should reveal themselves through microlensing against background stars. But where can we find large numbers of stars beyond the Milky Way's halo, yet still close enough to see them individually? The Large Magellanic Cloud. Unfortunately, Petrou did not publish her results, and in science the effect of not publishing is almost the same as not doing the work at all. Despite their insight, neither Refsdal nor Petrou have ever been fully recognised for their contributions.

The scientist who deserves the credit for the idea of using microlensing to search for MACHOs is Bohdan Paczynski – an amazingly prolific and exceedingly smart theoretician. Polish by birth, he has worked in Princeton for some time, at the Institute for Advanced Studies and at Princeton Observatory. In 1986 he developed a very simple and direct analysis of the microlensing process, including what microlensing would look like during an event, how long it would last based on different mass/distance combinations, and how many are likely to be seen over a given period of time. All this was published in a very simple, down-to-earth paper that finally prompted astronomers to think about gravitational microlensing as a way of searching for dark matter.

It just so happens that at the time Paczynski was working on this, Ken Freeman was visiting at the Institute of Advanced Studies. One day Paczynski came into Freeman's office and began talking about microlensing as a way of finding dark matter in the Galactic halo. He had calculated that such a search would require monitoring in the order of a few million stars per night to detect a statistically useful number of microlensing events, and asked Freeman whether there was there any way one could conceivably monitor such huge numbers of stars simultaneously? Freeman's response was that it would be possible to conduct a search for MACHOs photographically – a technique that was in fact used by a team of French astronomers called EROS. The method involves using a Schmidt telescope to obtain a wide-field image of the Large Magellanic Cloud. On such a photograph, each square degree would include the images of around a million stars, and a single Schmidt telescope field could capture 10 million stars. The problem is in determining which of them change in brightness? To do this, each 15-inch square photographic plate would have to be analysed using a microdensitometer, and the brightness of each star image would be digitised and then compared with records to detect any changes. Even for a team of people, this seemed a very large task. To complicate matters, in the mid-1980s the photographic era was coming to an end, and high-powered microdensitometers were becoming increasingly scarce. Freeman's view was that this was technically feasible, but probably impractical. This was not good advice.

The idea remained dormant until one day in 1990, when Freeman received an e-mail from Dave Bennett, a post-doctoral researcher at Princeton, saying that he and his colleagues were considering a search for MACHOs through microlensing, and would Freeman and his colleagues at Mount Stromlo be interested in participating? Rather than search for microlensing events photographically, this team, under the leadership of Charles Alcock, of Lawrence Livermore National Laboratory, planned to conduct the search electronically, but to do this they needed two further ingredients.

First of all they needed a telescope of at least 1 metre aperture (although even this would not really be big enough) that could be dedicated to this search over an extended period of time. The other requirement was a large electronic imager based on an array of charge-coupled devices (CCDs). CCDs are now ubiquitous devices that act as the imaging element in everything from the Hubble Space Telescope to the family video camera.

At that time, CCDs normally used in astronomy were small, and were used to image small areas of the sky. But although there are millions of stars in the Large Magellanic Cloud, the chances of a MACHO lining up between us and any one of those stars would be tiny. Some MACHO events were expected to last for less than a day, and so monitoring at less than this frequency runs the risk of missing an event. Furthermore, to improve the chances of catching a MACHO microlensing event in the act, the astronomers would need to monitor millions of stars every night. To be able to do this, the required CCD would need to be of a size the astronomical world had never seen. What Alcock was proposing seemed risky at best. Microlensing had been predicted theoretically, and Paczynski had written his paper on it, so astronomers knew the poor statistical odds of success. In short, there was no obvious reason why it should not work – and yet microlensing had never been seen. To invest such huge sums of money into an untried project would be quite a risk, but they went ahead with it.

The Lawrence Livermore National Laboratory is not far from San Francisco, and Alcock had a close association with the University of California at Berkley, where one of the big physics groups at the time was the Center for Particle Astrophysics (CfPA) headed by Bernard Sadoulet. A major goal for the CfPA is to find exotic dark matter particles such as axions and WIMPs (see Chapter 12), and they were funded to build particle detectors for the task. The CfPA physicists also wanted to carry out a search for MACHOs – at least partly because the failure of a search for lumps of dark matter would strengthen the already strong case for research on exotic dark matter particles. Working at the CfPA was an exceptionally capable young experimental physicist, Chris Stubbs, who examined the problem of building an array of CCDs as a large area detector. His immediate response was 'This is easy!'; but some members of the team were not so sure, because Stubbs had had little experience in astronomy and working with CCDs. Nonetheless, to him such a device seemed not so difficult, and he went ahead and successfully built the necessary detector: two arrays each of four large ($2,000 \times 2,000$-pixel) CCDs capable of imaging 500,000 stars simultaneously. Such arrays are common enough now, but a decade ago they were a huge step forward in imaging technology.

Because the Large Magellanic Cloud is peppered with variable stars (which vary in brightness due to some intrinsic process) the astronomers needed some way of distinguishing between a variable star's change in brightness and a true MACHO event. Fortunately, there is a way. Variable stars do not change brightness the same way in every colour, and a plot of the changes will be different for red light than green light. A MACHO, on the other hand, will affect all of a star's light in the same way, so that any change will be the same for all colours. To Stubbs the solution was easy: record the images of 500,000 stars simultaneously, in two colours! Stubbs was a vital element of the MACHO project. Although all of the team members were crucial to the success of the programme, it is probably fair to say that if it were not for Stubbs' technical expertise, the MACHO project might not have seen the light of night.

But there was one ingredient still vital to the project: a telescope. This is where Mount Stromlo Observatory enters into the story. Sitting decrepit and disused in a sealed dome on the mountain was a 1.2-metre telescope with a long history; but thanks to the perseverance of Mount Stromlo astronomers and engineers it

Figure 9.1. The Great Melbourne Telescope in its original incarnation. Crippled by poor design and maintenance, it had to wait decades before it made any important contributions to the science of astronomy.

would soon make a major contribution to the search for dark matter. This was the Great Melbourne Telescope.

THE GREAT MELBOURNE TELESCOPE

The Great Melbourne Telescope was built by Grubb in Dublin in 1868, for the Victorian Government, and was the first of Australia's big telescopes. At the time it was the largest fully steerable telescope in the world, but it was plagued with mechanical and optical problems. Although it was deemed to be an excellent telescope when it was completed, trouble began immediately it reached Melbourne.

Firstly, the mirror was made of speculum metal – an alloy of copper and tin. This was outdated technology, having been superseded by silver-on-glass mirrors in the late 1850s. Moreover, a glass mirror can be recoated with silver or aluminium without affecting its figure, but a speculum-metal mirror requires an optical repolish, and it has a much lower reflectivity.

The Melbourne Telescope's huge mirror was given a coat of shellac to protect its surface during the four-month voyage from Dublin to Melbourne. Normally,

Figure 9.2. Restored and housed at Mount Stromlo Observatory, the Great Melbourne Telescope was used to search for MACHOs. Night after night, the telescope was used to gather the light of millions of stars, which was later analysed for signs of gravitational microlensing. The telescope was destroyed during the bushfires that swept through the observatory in 2002.

the shellac would have been removed with alcohol, but the astronomer in charge of the telescope mistakenly tried to remove the shellac with methylated spirit and water, and the once highly polished surface was severely degraded.

The original mirror was replaced by a second mirror, but the improvement was not great. Problems were made worse by a poorly designed mounting and an open lattice tube which caused the telescope to vibrate in the wind. Having a long focal length did not help either, since such a design is not ideal for long-exposure photography required to study fainter stars and nebulae. The telescope was even blamed for damaging the evolution of reflecting telescopes, which were soon to replace refractors as the world's primary research instruments.

In 1947 the telescope was sold cheaply to Mount Stromlo Observatory, near Canberra, where it was given a second life. It was fitted with a new modern mirror, and the tube was cut in half to accommodate the new optics. It became a very useful astrophysical telescope that was used extensively in the 1960s and

1970s for, amongst other things, some very innovative photometry on faint stars in the Magellanic Clouds, by Ben Gascoigne. It was also equipped with a scanning spectrophotometer, and several PhD theses were written at Mount Stromlo using data gathered with this instrument. But around 1979 the telescope had a major seizure, and one of the bearings welded solid. At that time, Mount Stromlo resources were being poured into a new 2.3-metre telescope at Siding Spring Observatory, near Coonabarabran, and without funds to repair the Great Melbourne Telescope it once again lay idle and alone in the dark.

Working at Mount Stromlo at the time that Bennett approached Freeman, was Peter Quinn who, like Freeman, worked in the field of galactic dynamics. Together, Quinn and Freeman approached the then Director of Mount Stromlo and Siding Spring Observatories, the late Alex Rodgers. Rodgers was an adventurous character, and he decided to support the project – a very courageous decision, considering its potential for failure. This was untried science, and the MACHO project would never have been possible had it not been for Rodgers' courage in taking full responsibility and his decision to channel a significant fraction of Mount Stromlo Observatory's resources into the project. This move was not entirely popular among some of the observatory staff; nonetheless, the MACHO project was born.

SOFTWARE DEVELOPMENT

Analysing the light from 10 million stars every night called for a data pipeline of a size that had never before been attempted for ground-based astronomy. An enormous amount of work was required in preparing the software for the MACHO project. Tim Axelrod – an astronomer with a strong technical background – used to work at Livermore, but in 1991 he moved to Mount Stromlo with his wife, Robyn Allsman, who is a UNIX systems expert. It was intended as a year's sabbatical, but they liked it so much that they decided to sell up and move there permanently. Axelrod worked for many years on the staff of Mount Stromlo, and his wife was a senior person at the supercomputer installation of the Australian National University. Tim and Peter Quinn were responsible for the online software, and David Bennett wrote the image analysis software used for stellar photometry. Since there was no commercial database that was satisfactory for the purpose, the team wrote its own. The final database has about 20 million stars, with many hundreds of measurements per star.

THE FIRST MACHO EVENT

Observations began in mid-1992. The astronomers were looking for a variation in the brightness of a star with specific characteristics. First of all, it would have to vary in brightness in exactly the same way in both red and green light (as described earlier). This would rule out any physical changes within the star itself,

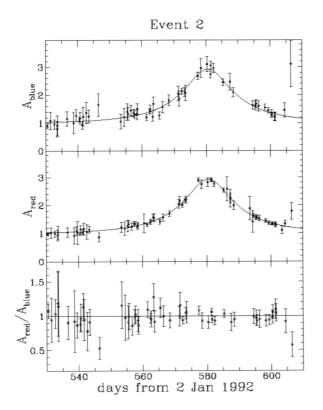

Figure 9.3. A MACHO event. Gravitational microlensing as a MACHO passes between Earth and a star causes the brightness of the star to be temporarily magnified.

or of the appearance or disappearance of a second star, as can occur in a binary system. Secondly, the light curve – a plot of a star's brightness over time – would have to be symmetrical before and after the peak of the brightening.

It was not long before the first event was seen. It was spotted by MACHO team members in the US, who were monitoring the stellar light curves as they came in, and the news soon spread world-wide. So great was the excitement about the team's discovery that it made the front cover of the journal *Nature*. It was dubbed the Gold Plated Event, because it was the first certain major microlensing event seen for stars in the Large Magellanic Cloud. Several smaller events were suspected, but the GPE was the first undeniable microlensing event. The amplification of the starlight was a factor of seven, and the light curve was beautifully symmetrical and achromatic, exactly as predicted by theory. It was inconceivable that it could be anything other than microlensing – probably by a MACHO passing between the star in the LMC and the Earth.

LOOKING AT THE CENTRE OF THE GALAXY

The Large Magellanic Cloud is high in the night sky only during the southern hemisphere's summer months. In winter, observations were conducted towards the centre of the Galaxy. Looking out towards the Magellanic Clouds is rather different from looking at the centre of the Galaxy. We look out through the plane of the Galaxy, then through the halo, and into the LMC. In this direction the MACHO team sees about three microlensing events every year; but in contrast, every year they see about fifty microlensing events towards the centre of the Galaxy. These events probably have nothing to do with dark matter, since they are most probably just stars microlensing other stars. But there are many more microlensing events seen towards the inner Galaxy than the team expected, and this is still not understood. It may be that there is something wrong with the Galactic models that we use, or that there is rather more dark matter towards the inner part than we previously thought. Although this seems unlikely, it is possible that these microlensing events towards the core of the Galaxy are, after all, dark matter related.

RESULTS

At the time of writing (2005), the team has completed its observations and is involved in analysing seven years of data. If the events that have been seen in the direction of the Magellanic Clouds are due to dark matter objects in the halo, then they are consistent with the objects being half the mass of the Sun, and there seem to be less than half as many as we need to describe the halo. Now, both of these numbers have pretty big uncertainties in them still, though those uncertainties will be smaller once the team has completed its study. The mass of the objects could range from 0.3 to 0.8 solar masses, and while this does not tell astronomers what MACHOs are, it does seem to rule out brown dwarfs, which are about a tenth of a solar mass. White dwarfs – faint remnants of dead stars with masses of about 0.6 solar masses – are an interesting possibility for the MACHOS. Several teams are now busy looking for this population of white dwarfs, using other techniques such as detecting them by their apparent motion in the sky.

PROBLEMS AND UNCERTAINTIES

Still, not all astronomers agree that the MACHO project has detected MACHOs in the halo of our Galaxy. The stars of the Magellanic Clouds, they argue, do not lie in a flat plane, but are instead spread out along the line of sight. Some microlensing events could therefore occur as Magellanic Cloud stars pass in front of each other. Further ammunition to these critics is the fact that of only two events witnessed in the Small Magellanic Cloud, at least one has turned out to be a binary object; and when a binary object is involved, it is sometimes possible to estimate the distance of the lens.

With simple lensing of a single star by a single object, the really important parameter is the duration of the microlensing event. This depends on the mass of the lens, where it lies relative to the star, and how fast it is moving. We think we know the probability distributions of the lens locations and the lens velocities (this just means the chance of finding the lens in some location and moving at some speed), so the duration provides a statistical measure of the mass of the lens. On the other hand, the amplification reveals only how close the lensing object lies to the line between you and the star – and that is not so useful.

For a binary lens, the whole lensing calculation is much more complicated, but it is sometimes possible to estimate the actual distance of the lens. It happens that they lie fairly close to the Magellanic Clouds.

The chance of seeing a lensing event is greatest when the lens is about half way between us and the star. When the lens and star are close together, the chances of seeing an event get very small. This is why most people do not believe that the MACHO events seen for the Large Magellanic Cloud could be produced by LMC stars lensing other LMC stars. The LMC is a fairly flat galaxy seen close to face-on, so the lenses and the stars being lensed would have to be fairly close together along the line of sight. This is rather worrying in the case of the LMC. On the other hand, it is much less of a worry for the SMC, because we know that the SMC is not at all flat, and that is is strung out along the line of sight, so we would expect to find SMC stars lensing other SMC stars.

MAGELLANIC STREAM DEBRIS

Of greater concern is the possibility that the microlensing events do not indicate dark matter at all, but rather that we are seeing microlensing by shreds of tidal material that might lie between us and the Magellanic Clouds. So far, only gas has been seen in the Magellanic Stream (as discussed in Chapter 5), despite decades of searching for stars. But just suppose stars are there, say 10–15 kiloparsecs in front of the Large Magellanic Cloud – or even behind it would suffice – in which case the number of microlensing events would be about what we see now. Astronomers have been looking for this tidal debris in the MACHO database photometry, and also in independent projects. Debris between us and the LMC would show up, for example, as variable stars that are slightly brighter than the same kind of variable star in the LMC itself. So far, there is no sign of this tidal debris, but it is not yet an entirely closed question.

At this time, most astronomers believe that the MACHO project has detected lensing objects in the halo of the Galaxy, but not enough of them – 20% at most – to account for all of the dark matter that is known to be out there.

VARIABLE STARS, IF NOTHING ELSE

Even if microlensing had turned out to be non-existent, a very positive aspect of the MACHO project would remain. Monitoring 20 million stars nightly resulted in an enormous amount of data on variable stars, and many papers have been published. Many other people have since used the database for projects that have nothing to do with dark matter, and a great deal of science about stars has come out of the project.

SEARCHING FOR EXTRASOLAR PLANETS

During the winter, when the MACHO project was looking at the heart of the Galaxy, it was seeing on average almost an event per night. As we said earlier, what is seen towards the centre of the Galaxy is mainly self-lensing events by stars. Now, if the star responsible for the lensing has a faint companion or planet, in principle the sign of the planet can be seen in the lensing curve. It produces a fairly complicated light curve with a few kinks and bumps that are rather short-lived – perhaps an hour or so. So, one of the ways people go planet searching (which is currently a very hot topic) is to use microlensing. Once it is known which star is being microlensed it can be studied pretty intensively – perhaps even with a world-wide network of telescopes constantly monitoring it. This would mean that as one observatory – say Mount Stromlo – experiences sunrise, another observatory further west – say in South Africa – can take over and continue monitoring the suspect star. And there are a couple of groups doing this. An automated alerting system interrogates the database all the time, updating the astronomers on any stars behaving in an interesting way. When the brightening of a star is detected and it is not a known variable star, numerous e-mails are sent to tell astronomers which star it is, where it is, and what it is doing. This is updated on a daily basis, providing the astronomers with a chance to swing into action. Instead of just checking on the star once or twice a night the way we do, they just sit on the star constantly, monitoring it for the signals of a planet. The results have been promising, and include the first detection of a planet orbiting a binary star. Furthermore, this technique is the only known method of detecting Earth-sized planets orbiting Sun-like stars – planets that may be suitable for life as we know it.

THE FUTURE OF MACHO

To derive much more out of the original MACHO project, it would have to run for a great deal longer. As more events are observed, the errors reduce as the square root of the number of events. For example, to halve the statistical uncertainties, four times the number of events are required, so that the experiment needs to be repeated three times – another 24 years! And so the

MACHO astronomers have called it a night. The MACHO project was completed in 1999 after detecting many microlensing events towards the Magellanic Clouds and the bulge of our Galaxy. The equipment was turned to searching for objects closer to home: trans-Neptunian objects in the outer reaches of the Solar System.

In 2003, bushfires swept through the countryside around Mount Stromlo and completely destroyed the observatory. Not only was this a terrible loss of an historic research facility, but irreplaceable historical records and equipment, and teaching facilities, were also destroyed. The fire continued on towards Canberra, engulfing homes and lives. Despite such terrible losses, the data from the MACHO project were safely stored on campus at the Australian National University and in the US.

The search continues. Other groups are also looking for MACHOs, such as the EROS project (mentioned earlier). EROS and a Polish–American team are both now operating with 1-metre telescopes and large modern detectors in Chile, where the climate and the astronomical seeing are much better than at Mount Stromlo.

10

What can the matter be?

BARYONIC DARK MATTER: WHY IT IS SUSPECTED

Until recently there was considerable encouragement that dark matter was baryonic. The argument was that we had a number from the BBNS that was somewhere around $\Omega_b = 0.04$; that is, BBNS predicted that the amount of baryonic matter made up about 4% of the critical density. Visible matter comprises only about a fifth of that amount, but when the dark halo matter is added they constitute about 4% of the critical density as predicted. This encouraged astronomers to consider, until about a year ago, that there were many baryons out there, left over from the BBNS, that had not yet been seen. But then it was realised that there is an enormous amount of hot gas around, and that it accounts for much of the baryon budget allowed by BBNS.

These days, when the visible baryonic matter – stars, gas, everything – is totalled, the figure is probably about right both at $z = 3$ and at the present. The errors involved are still pretty high, however, but the encouragement to think that dark matter is baryonic has evaporated. Nonetheless, it is still worth exploring the baryonic as well as non-baryonic options for dark matter. In doing so, not only do astronomers need to explain what a dark matter candidate is; they also have to explain how it got there and, ideally, provide proof that the candidate is what they say it is.

FAINT STARS

With baryonic options, objects we can more or less exclude with certainty include faint stars. Astronomers thought at one point that the Universe might contain many dwarf stars – not brown dwarfs (which we will consider shortly), just stars that are very faint – but the Hubble Space Telescope conducted a series of long observations in the Hubble Deep Field so that we now have a pretty good census of the population of very faint stars; and there are nowhere near enough of them. At this stage faint but otherwise normal stars are out of the picture. Moreover, the MACHO project and the French EROS experiment has now excluded anything with a mass between 0.0000001 and 0.1 solar mass. If such objects existed as compact lumps, we would have seen them.

SMALL HYDROGEN SNOWBALLS

There are a few other candidates that we can mention and almost immediately exclude; for example, small hydrogen snowballs with a mass less than 0.0000001 solar mass. However, theoretical work has shown that such objects would quickly evaporate in the background radiation of the Universe. Another possibility is that there might be very tiny objects with the mass of an average house-brick; but if there were enough of them we would experience a constant rain of high-velocity meteors streaming through the Earth's skies. Clearly, we do not see such objects falling from the heavens, and so every time you walk out under a clear, meteorless night you are participating in a dark matter investigation, albeit with constantly negative results.

MASSIVE BLACK HOLES

There is yet another strange beast that might contribute to the dark matter problem: black holes. All matter in the Universe has gravitational attraction, and as we have seen, gravitating objects can not only pull you, aircraft and rockets toward them, but also light. This phenomenon was first demonstrated by Eddington in 1919, as proof of Einstein's General Relativity, and has since been demonstrated in spectacular fashion in the form of gravitational lenses. But all this is merely the bending of light. What if it could be trapped forever? Imagine an object so dense that nothing – not even light – can escape its gravitational grasp. To an outside observer, they look like a hole in space where everything – all matter and light – falls in and disappears from our Universe forever. Could such bizarre objects account for dark matter? After all, many of them are massive, and all are certainly dark! Many galaxies are now known to have black holes at their centres, and the Milky Way is no exception. This is quite a different form of dark matter, and requires entirely different techniques for finding it. A black hole is a massive object at the opposite extreme of halo dark matter, right at the centre of our Galaxy. Its mass of about 3 million solar masses seems enormous, but in the scheme of things it is not that much; only about the mass of a large globular cluster. What is interesting is that black holes are dark, and it is believed quasars and other strong radio sources such as active galactic nuclei are powered by them. The power of a black hole is produced by matter falling into it – a process that releases tremendous amounts of energy. This gravitational energy has to go somewhere, and there are ways of converting it into radio waves and light.

As yet it has been very difficult to find a black hole in an active galaxy. The techniques used are compromised because the nucleus is active, and it is not possible to determine what is causing all the commotion! So, most black holes that have been discovered have been found in relatively quiet galaxies. These galaxies may at one time have been active, but due to their current lack of activity the black holes can be seen. At the centre of our Galaxy, for example, is a radio source called Sgr A* (Sagittarius A star). The discovery of this object prompted a search for a black hole, by which the motions of stars at the very centre of the Galaxy were observed – which is very difficult, because the centre is obscured by dust.

In the infrared, however, the problems caused by dust are greatly decreased, and in recent years two groups have been studying the motions of stars very near the Galactic centre by studying them in the infrared. The really interesting information has arisen from their attempts to measure their transverse motions. The closer a star is to the centre of the Galaxy, the faster it has to move to escape being sucked into the black hole. So, in order to determine the size of the black hole, astronomers set about trying the measure the velocities of the Galaxy's innermost stars.

The techniques used for this study are elegant. The visible motions of stars here are tiny – so small, in fact, that due to the turbulence of the Earth's atmosphere they are almost impossible to see. Stars twinkle in the night sky because their light is distorted, or blurred, by the Earth's atmosphere. While they appear sharp and bright to the naked eye, to modern astronomers using state-of-the-art telescopes it is like standing on the edge of a swimming pool and trying to read a book at the bottom of the water. The pages can be seen, but the words are blurry. As well as blurring a star's image, the Earth's atmosphere moves it around in a random and unpredictable manner, which causes it to dance around in the field of view of the telescope. However, astronomers have managed to obviate this problem. One method is to use 'adaptive optics', with which a ground-based telescope can produce images of the same resolution as those produced by the Hubble Space Telescope. Adaptive optics involves a system of tiltable and deformable mirrors that counteract the effects of the Earth's atmosphere. In simple terms, every time part of the image moves, the corresponding part of the mirror moves accordingly to keep it focused sharply and cleanly.

An alternative technique is speckle interferometry, used by an American group. This involves taking very short (50-millisecond) exposures with a very large telescope. Each image of the star is broken up into lots of little bright speckles by the atmospheric turbulence, and each one of the speckles has a size that corresponds to the physical limit to the resolving power of the telescope, even if it were in space. With a large amount of processing, the astronomers can reconstruct an image that contains some of the high-resolution details that would be seen with the Hubble Space Telescope. It is not quite as elegant as adaptive optics, but it is much cheaper.

These two methods have been independently used to measure the motions of the same stars near the heart of the Galaxy, and they have achieved almost identical results. The velocities of the stars are unbelievably big: up to 2,000 km/sec. Compare this with the leisurely 220 km/sec motion of the Sun two thirds from the centre of the Galaxy. The mass of the central black hole needed to generate these velocities is an enormous 2,700,000 solar masses. This discovery represents a brilliant piece of observational work that caused great excitement when the news broke in 1997.

This idea leads to another aspect of dark matter that is quite different from anything we have discussed so far. No-one knows, for example, just how black holes form. They probably formed out of the gas during the early stages of the Galaxy's formation, rather than from dark matter, and at some point they simply became too dense to hold themselves up and fell in on themselves. Although the formation of black holes is not well understood, they occur in the centres of most galaxies that have central bulges, like our Milky Way.

Although we have so far discussed only the black hole at the centre of our Galaxy, there are several ways that objects of a million solar masses could form in the early Universe and then become black holes, not necessarily at the centres of galaxies. Astronomers have suspected for a while that such million-solar-mass black holes might make an attractive form of dark matter, and we could imagine that the dark halo of our Galaxy consists of about a million of these objects. Whether or not they are referred to as baryonic is a matter of taste, but they could make a formidable contribution to the mass of the Universe.

Such an idea would also help solve another problem. If we had a halo full of million-solar-mass objects, they would occasionally and randomly hurtle through the disk of the Galaxy and disturb the otherwise quiet lives of the stars there. Sure enough, observations of stars in the Galactic disk reveal that the older stars are more energetic than the younger ones – which might suggest that they have been there longer and have been stirred up more often by repeated encounters with massive black holes from the Galactic halo. It has always been rather a puzzle to find enough energy to drive these older stars to the observed velocities. While there are other ways to stir up disk stars – such as stars interacting with gas clouds, spiral arms, and so on – black holes are an attractive proposition. But there is a problem. Although million-solar-mass black holes would provide a good solution for a large galaxy like ours, for smaller galaxies they would be very disruptive. They would contribute far too much random stellar motion, and the disks would appear to puff up like so much dust disturbed by a falling stone. Such massive black holes would ultimately tear smaller galaxies apart. And there are plenty of smaller disk galaxies that are clearly still very flat disks in one piece, yet contain huge amounts of dark matter. Even in the case of the Milky Way such objects would leave their mark by smashing up the globular clusters that occupy the halo, and yet we see plenty of globular clusters surrounding our own and other galaxies. Massive black holes have therefore now been ruled out as probable candidates for the dark matter.

SMALL BLACK HOLES

It might be possible to still have smaller black holes – say of 1,000 solar masses – and at this point it is difficult to exclude them on observational grounds. For example, the MACHO experiment would not have detected them, because the typical time for the brightening and fading of a star lensed by a black hole of this mass is much too long: about fifteen years. In any case, there would still need to be an explanation about how such objects could form in the first place.

SMALL DENSE CLOUDS

One baryonic candidate that is still in favour with some astronomers is compact, cold, dense gas clouds that occupy the halo. If these clouds exist, they would have to be typically 0.001 the mass of the Sun – similar to Jupiter's mass – otherwise there would be very little preventing them from collapsing in on themselves to form stars. Furthermore, there must be some reason why they do not collide with each other and slowly accumulate enough mass to again form stars. And since none have been seen, they must be very hard to detect. This means that they are not only dark (they have no stars to light them up) and cold, but they must be small – say around a fraction of a parsec. But why small? If they were large they would have already revealed themselves to radio astronomers by absorbing radiation from distant sources such as quasars. These clouds are therefore very difficult to find.

From the MACHO project we can say that there are objects out in the halo around 0.5 solar mass. A very natural candidate that falls into this mass range are old – hence very dim – white dwarfs. Recall that white dwarfs are the dying remains of stars about the mass of the Sun, shrunk to spheres about the size of Earth. They have the disadvantage that they make an awful mess. In the process of becoming a white dwarf, a star will produce a lot of carbon and nitrogen. To make a 0.5-solar-mass white dwarf, the process must begin with something with a mass of probably 2 or 3 solar masses, and 2 solar masses of enriched material is recycled into the interstellar medium. It would be like having a whole population of diesel engines in the halo, all spewing out carbon. But there is simply not that much carbon out there. Yes, the enriched material might have been removed in some way; but so far, cleaning up the Galactic halo involves mechanisms that are so complicated as to be improbable.

Another argument against white dwarfs as a source of dark matter is the question: when were they made? Forming white dwarfs is a fairly inefficient way to make dark matter, when it is considered how much mass it has to account for. The process of star formation, evolution and formation of a white dwarf ties up only about a quarter of the original stellar material as a white dwarf, while the rest is thrown back into space. To have 90% of the mass of a galaxy tied up in this form, the stellar evolution process needed to create a white dwarf would have to have been completed many times. All this takes time, and it would take a fair fraction of the life of the Universe for a sufficient number of cycles of star formation and stellar evolution to take place. Astronomers continue to debate the merits of white dwarfs as the repository of the Universe's dark matter, but the idea has its problems.

Massive stars can create two other kinds of dark matter candidates: neutron stars of around the mass of the Sun, and black holes in the 20–100 solar mass range. Both would be the cores of massive stars that have exploded as supernovae. But again, the process of creating either a neutron star or a black hole creates a lot of enriched gas which is not seen.

If forming white dwarfs, neutron stars or black holes is the answer to the dark matter problem, it is not clear when this would have happened, but it could have been during a pre-galactic episode of star formation. Long before the galaxies formed, the Universe may have undergone a big burst of star formation. There are all sorts of arguments against this, however, concerned with how much infrared radiation would be seen in the background, and so on. So, the idea does not appear all that healthy.

BROWN DWARFS

One candidate that was very popular before MACHO was brown dwarfs – objects which were not massive enough to start nuclear burning and become stars. Such objects would be in the mass range of 0.001–0.1 solar mass. Again, such objects could have been created before the galaxies appeared. They are also much cleaner! If there had been a big burst of brown dwarf creation in the early Universe, there would be no evidence that it had happened. Once formed, they would behave just like dark matter.

At the time that the MACHO project was planned, brown dwarfs were the favourite candidate, so MACHO was designed to find objects around 0.1 solar mass – brown dwarf size – simply because that is what the astronomers were expecting. Brown dwarfs have been found in small numbers in the disk of our Galaxy. For a while it was hoped that the humble brown dwarf would also make a significant contribution to dark matter in the Galaxy's halo; while not massive, there could possibly have been lots of them. This now seems unlikely, however. From MACHO, it seems pretty clear that they are not out there in the halo in sufficient numbers. Chris Tinney – a research astronomer at the Anglo-Australian Observatory – has searched directly for brown dwarfs among the data produced by the Deep Near Infrared Survey (DENIS), and other astronomers have searched using the Two Micron All Sky Survey (2MASS). While he remains optimistic that DENIS and 2MASS will find lots of brown dwarfs, Tinney points out that the numbers are not sufficient to integrate up to a significant proportion of dark matter.

PRIMORDIAL BLACK HOLES

If the MACHO project is indeed finding dark MACHOs in the halo, with masses around 0.5 solar mass, then there is another attractive possibility: primordial black holes. They present an attractive option for the one simple reason that it is such a clean solution. The only viable alternative to a dark matter explanation for the MACHO events is white dwarfs; but as we have said, this explanation involves the problem of all that extra chemically enriched material in the halo which simply is not there. Primordial black holes, on the other hand, can have the right mass, and are self-contained and clean.

There is then the problem of how to make these things. The time to make primordial black holes in the 0.5-solar-mass range is microseconds after the Big Bang, and astronomers are currently divided as to whether it is possible to do so. The aim is to force an object to collapse into a black hole, and the size of the object depends on the event horizon at that time; that is, how far light can travel at that stage in the Universe's history. During the very early stages of the Universe the light travel distance was very small, so only very small objects could be created. As time went on, bigger objects could be created; and at about 10 microseconds after the Big Bang, black holes around 0.5–1 solar mass could be created. This was also the time when the Universe went through one of its so-called 'phase transitions', during which there were probably numerous fluctuations, analogous to steam condensing into liquid water. At that time the Universe changed from being made mainly of quarks to one made of hadrons. (We will discuss this in more detail in the next two chapters.) It just so happens that the quark–hadron phase transition was just about at the time of the creation of 0.5-solar-mass black holes. Primordial black holes cannot really be called baryonic. It happened so early in the Universe that the Universe itself did not yet contain baryonic material. Certainly it happened before BBNS, so matter was locked up in these primordial black holes long before it had a chance to react with atomic nuclei.

BETWEEN THE GALAXIES

Recently an entirely unsuspected source of stellar baryonic matter was discovered in the vast expanses between the galaxies. It has always been assumed that the space between the galaxies was, save for dark matter and hot gas, more or less empty. As it turns out, there is indeed stellar matter there; but it is not dark, just extremely faint. Freeman and a team of collaborators from Europe and the US discovered this previously neglected source of baryonic material in 1996. When you look at a cluster like the Virgo or Coma Cluster, you see the galaxies and the gas, but between the galaxies there are lots of stars. Although in the Coma Cluster these individual intracluster stars are invisible because the cluster is so far away, nonetheless they reveal themselves as a very diffuse distribution of light. In the much closer Virgo Cluster, individual giant stars can just be distinguished with the Hubble Space Telescope.

HOW TO FIND INTRACLUSTER STARS

From the ground we can see planetary nebulae that belong to the intracluster stellar population. As discussed in Chapter 5, planetary nebulae are brighter at a specific wavelength in the blue–green part of the spectrum than anything else around them, and are therefore efficient markers of nearby matter. To find a planetary nebula, an image of a preselected area of a cluster must first be obtained. But this is no ordinary image, as the telescope is fitted with an interference filter, which excludes all the light except the light emitted by the planetary nebulae. (Specifically, it is the oxygen line, [OIII]. Amateur astronomers use similar though much simpler [OIII] filters to enhance the visual appearance of planetary nebulae.) A second image is then obtained – this time using a filter that allows through everything except the [OIII] line. When the two images are compared, many objects appear in the image taken through the [OIII] filter that are invisible in the image that excludes [OIII]. As might be expected, ordinary stars are visible in both images, but planetary nebulae – being very faint in wavelengths other than [OIII] – show up only in the [OIII] image. Using a 4-metre telescope it takes about five hours to build up an image that reveals a sufficient number of these objects in a cluster like the Virgo Cluster.

Astronomers have been looking for planetary nebulae in galaxies for quite a while, because they are one of the main distance-measuring tools. The brightness of the brightest planetary nebulae is fairly constant from galaxy to galaxy, and so they can be used as standard candles. The term 'standard candle' derives from the analogy of the brightness of a candle held at arms length compared with the brightness as it is taken further and further away. Since the brightness will reduce as the square of the distance, comparison of the brightness of a candle some distance away with its known brightness when nearby yields the distance to the candle. Similarly, by measuring the brightness of a planetary nebula in a galaxy and comparing it with the brightness it would appear nearby, the distance to the planetary nebula, and the galaxy in which it resides, can be determined.

However, Freeman and his colleagues were initially interested in planetary nebulae for another reason: studying the dynamics of galaxies. Although difficult to find, once a planetary nebula is in your sights you can measure its redshift precisely from the location of the prominent [OIII] line. Measurement of a hundred or so planetary nebulae around a galaxy allows measurement of the galaxy's gravitational field – a process similar to that of measuring the random motions of stars, described earlier. They are therefore ideal objects for measuring the distribution of dark matter in elliptical galaxies.

Another application of planetary nebulae arose by accident, when Freeman and his colleagues found some in the space between the galaxies of the Virgo Cluster. After looking at the spectra of these isolated objects, they confirmed that the planetary nebulae were part of the Virgo Cluster and not strange foreground objects. Many astronomers have since undertaken similar searches, and it turns out that there are very large numbers of planetary nebulae in the huge voids between the galaxies. In the Coma Cluster there is about as much starlight between the galaxies in clusters as there is in the galaxies themselves. It seems a whole reservoir of baryonic matter had been overlooked! But how did they get there? Why, when at least half of the stars in the Universe are huddled together in galaxies, do these loners wander the vast expanses of space between the galaxies? The answer may be that they were torn from their host galaxies by a process called harassment. Early in the story of dark matter we saw that clusters of galaxies are dynamic places – more so for the existence of dark matter. And so, as the galaxies career past each other over the aeons, their gravitational fields shake stars loose. The tidal field of the cluster then pulls the stars away from the outer layers and sends them flying into space. This is by no means certain, but a considerable amount of recent computational modelling indicates that it is the most probable explanation. On the other hand, it is possible that there may have been an era of star formation before the galaxies formed. A number of problems remain unanswered about stars between the galaxies; for example, how they are distributed in space. If they really are torn from galaxies they could be expected to form ribbons arcing across the cluster, and there would be places where there are no stars at all. If they are due to some previously unknown period of star formation, then their distribution should be more even. Just as important is the way these intracluster stars are moving about between the galaxies. A considerable amount of observing time will be required to answer these questions.

Freeman and his colleagues carry out such observations on 4- and 8-metre class telescopes, at every opportunity. To speed up the measurements they have built a dedicated instrument called a planetary nebula spectrograph, which is mounted on the 4-metre William Herschel Telescope at La Palma, in the Canary Islands. With simultaneous imaging and spectroscopy, the spectrograph is used primarily for finding and measuring the velocities of planetary nebulae around elliptical galaxies, for studies in the dynamics of dark matter.

However, more extensive studies of planetary nebulae in the intracluster medium require telescopes larger than 4 metres. Despite their excellent performance, the light-gathering power of 4-metre telescopes limits them, for planetary nebulae, to distances of about 20 megaparsecs, which includes clusters such as Virgo, Ursa Major, Fornax, and so on. But there are only a few clusters within this distance, and none of them are particularly rich. Astronomers would like to reach out to the Coma cluster, for example, and this has now become possible using the light-gathering power of 8-metre telescopes such as the Japanese Subaru telescope on Mauna Kea, in Hawaii. The Coma Cluster is dense, and if the ideas about stripping stars are correct, then we would expect a much higher fraction of stripped stars in these clusters simply because the harassment process would be more efficient. And that is what we find: about half the starlight in the Coma cluster lies in the intracluster space, between the galaxies.

INTRACLUSTER GAS

As well as stars, a lot of gas exists between the galaxies. The motion of the galaxies through this gas heats it up so much that it emits X-rays, which can be studied using satellites. But there is an unsolved puzzle in this X-ray emission. Chemical element lines can be seen in the X-ray spectra, and from studying the lines due to iron, astronomers have been able to measure how much iron exists in the hot gas. The proportion of iron to hydrogen is much higher than expected: the intracluster gas has been somehow polluted by chemically enriched material. It has never really been understood why this is so. One possible reason is that some of the chemically enriched gas produced by supernovae exploding in the galaxies can escape from the galaxies and pass out into the intracluster gas. This material has to escape from the surrounding gas and gravitational field in the galaxies, which is not so easy. But when a star is floating in free space there is nothing holding back the material. Those stars therefore have probably contributed a lot of chemically enriched gas to the intracluster medium – probably much more than the stars in the surrounding galaxies.

MILGROM'S ALTERNATIVE THEORY OF GRAVITY

Now for a strange twist in the search for dark matter. What if it is all an illusion? What if dark matter does not really exist? As we have repeatedly seen, one of the primary pieces of evidence of dark matter is the rotation curves of galaxies: stars travelling too fast to be held in place by the visible matter alone. When we use rotation curves to measure mass, we are really measuring the acceleration of gas moving around the galaxy in circles. Galaxies are very large, so this acceleration is very small – much smaller than the accelerations that Newton was considering when he established his laws. So there is always the possibility that Newton's laws do not apply on galactic scales. If you assume that most astronomers disregarded this possibility, you would be right. After all, the Newton–Einstein laws of gravity have been working extremely well for hundreds of years, and have allowed the discovery of everything from planets and dwarf stars to gravitational lenses. These are cherished laws. To suggest that they could be wrong is a potentially dangerous act (for the claimant). Nonetheless, it is an alternative to the dark matter mystery that we need to mention. So at this point it is worth looking at the possibility that rather than the existence of dark matter being the explanation for these flattening and rising rotation curves, Newton may have been wrong for large scales.

One scientist that has made such a radical claim is Mordehai Milgrom, of the Weizmann Institute of Science in Rehovot, Israel. Working at the Condensed Matter Physics Department, Milgrom has carried out a great deal of work on the possibility that at low accelerations Newton's laws of gravity break down. You might recall from high-school physics that the more massive an object is, the more gravity it has. In fact, the two are directly proportional: the gravity is proportional to the mass and inversely proportional to the square of the distance from the mass. Gravity is an acceleration: the longer an object falls, the more it accelerates, and so on. Milgrom proposed that when the acceleration is very low it becomes proportional to the square root of the central mass and inversely proportional to the distance (not the square of the distance). Although no-one knows if this simple assumption is correct, it automatically produces flat rotation curves in the outer parts of galaxies, without any need for dark matter. He has applied his alternative theory – known as Modified Nonrelativistic Dynamics (MOND) – to galactic, cluster and supercluster scales. It works well, and so far no-one has managed to find any way to disprove it. According to Milgrom, when you look at the Universe this way there is no need for dark matter at all, as it simply does not exist.

Many other scientists have proposed alternative theories of gravitation that do away with the need for dark matter; but at present, Milgrom's is the most widely discussed. Many astronomers are not ready to consider it seriously, but others are becoming more impressed at how well his simple assumption actually works in practice. But if Newton and Einstein were correct about gravity, if BBNS and all the observations of baryonic matter are correct and accurate, and if we can still only find a fraction of the gravitating matter in the Universe, we are left with one option: if it is not baryonic, it must be something else...

11

Exploring exotica: neutrinos

Welcome to the Universe of non-baryonic matter. This is a Universe made of particles far stranger than the ordinary matter that makes up the pages of this book, the air you breath, or even you. So strange is this non-baryonic world that it may be made of particles that have not yet been detected – that are so far merely hypothetical. Others are known to exist, but the question remains: are there are enough of them to make up the dark portion of the Universe, and is there enough of it to explain the behaviour of the visible Universe?

WHY NON-BARYONIC DARK MATTER IS SUSPECTED

A major problem facing dark matter researchers is that particles of non-baryonic matter are far harder to detect than anything covered so far. While they seem to have a gravitational influence over stars and galaxies, they generally interact only feebly with the materials from which conventional particle detectors are made. For example, one popular dark matter candidate – the neutrino – can travel through this book, you, the entire planet, and much, much more, without bouncing off or otherwise interacting with a single grain of sand, a molecule, or any other particle you care to imagine. Despite this antisocial behaviour, neutrinos may have mass; and if they have mass they can influence, through the force of gravity, the behaviour of anything else that has mass. And if there are enough of them – or any of the other particles we are about to encounter – they may be responsible for the unexplained behaviour of the Universe.

So now there are two problems with non-baryonic particles that have to be overcome: how to detect them, and how to measure their mass. But before we address such important questions, it is logical to ask exactly why such intangible material – non-baryonic matter as opposed to some other solution to the dark matter problem – is suspected in the first place. Well, for one thing, the same theories that predict the abundance of baryonic matter, Big Bang Nucleosynthesis (BBNS) along with observations and formation theories of galaxy clusters, also predict the existence of non-baryonic matter. In Chapter 8 we saw how BBNS predicts a certain amount of baryonic matter, and that this prediction ties in reasonably well with observations of a variety of phenomena. There is a concordance between the measurements of these abundances and their calculation, provided the amount of baryonic matter is within a certain range. We have also seen that the amount of matter needed to hold clusters of galaxies together is much higher than either the predicted or observed amount of baryonic matter. The amount of matter needed to explain the dynamics of clusters is about ten times the amount of baryonic matter, either predicted or observed. So, if dark matter is not made of baryonic matter, another option is that it consists of (not surprisingly) non-baryonic matter.

All non-baryonic material would have been left over from the Big Bang. As discussed in Chapter 8, immediately after the Big Bang the Universe began expanding and cooling. Initially the entire Universe consisted of a soup of energy and matter, and particles of matter and energy interacted with each other in a variety of ways. For example, photons begat protons and neutrons and their antiparticles, and then these particle–antiparticle pairs annihilated one another to produce photons. It was not long, however, before this annihilation phase ended and baryonic matter was produced by the interaction of protons and neutrons. But the Universe was still opaque; that is, photons could not travel far without being soaked up by matter. Indeed, it was this relationship which drove the reactions that created the baryonic matter abundances we see today.

A pivotal moment was 380,000 years after the Big Bang. The Universe had finally cooled below a temperature of 2,967 K, and particles began to 'decouple': particles of matter were no longer soaking up every photon that came by, but were instead allowing them to go on their merry way. This stream of photons is still around today, visible across the entire sky as a glow of microwaves known as the cosmic microwave background (discussed in more detail in Chapter 13). It was a transition from an optically opaque Universe to a transparent Universe. The decoupling of neutrinos was similar. Yet while photons and neutrinos were decoupling, others particles such as electrons and muons disappeared from the plasma via particle–antiparticle annihilation. The important point is that at this stage the different types of particles retained their individual identities, eventually coalescing into both the Universe we see, and the one we do not see, today.

Not all the particles created in the Big Bang were baryonic. Many types of non-baryonic particle also formed at the same time, and although such particles have not yet been found – or at least have not yet been measured and studied with sufficient detail to ascertain their role – they undoubtedly have played, and continue to play, a major role in the formation and evolution of the Universe. The zoo of these exotic particles is vast indeed, but what is important for our story is that some of them may be massive enough, or plentiful enough, to conceivably make up the dark portion of the Universe.

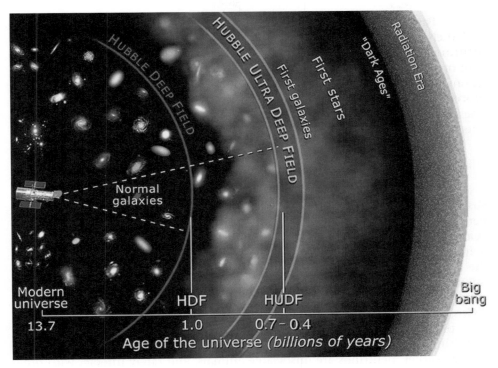

Normal
galaxies

Modern
universe

Big
bang

HDF

HUDF

13.7

1.0

0.7 – 0.4

Age of the universe *(billions of years)*

Figure 11.1. A history of the Universe has been slowly developed by observing objects at different distances and combining these discoveries with theory of how the Universe has evolved over time. Of critical importance to the evolution of the Universe is its initial composition.

CLASSES OF NON-BARYONIC DARK MATTER

From a cosmological perspective, all the dark matter in the Universe can be classified as either hot or cold. Hot dark matter (HDM) particles travel at relativistic velocities; that is, they move close to the speed of light. (They are not necessarily relativistic today, but simply needed to be relativistic at the appropriate time in the early Universe.) Everything else is called cold dark matter (CDM), consisting of particles that travel much slower than light, and including all of the baryonic matter we have discussed so far. In Chapter 13 we will explore the implications of HDM and CDM in the structure of the Universe, but for now we shall take a look at the most promising candidates for non-baryonic dark matter.

There are many generic types of candidate for non-baryonic dark matter, but amongst the plethora of possibilities three main types emerge. We have already met one of them briefly: the neutrino – a particle known to be produced inside stars and supernova explosions. Another type is the WIMP (Weakly Interacting

Massive Particle), which includes a large family of particles which at the moment are purely hypothetical. Examples are heavy neutrinos (different from ordinary neutrinos), neutralinos (including higgsinos, photinos and zinos), and gravitinos – all of which can exist in theory but are so far undiscovered. Finally there are axions, which, if they exist, would have been formed by somewhat different processes during the so-called strong interactions between subnuclear particles. These three classes of non-baryonic matter – neutrinos, WIMPs and axions – are alternatives for dark matter. We will now take a closer look at each of them.

NEUTRINOS

First, neutrinos – a form of hot dark matter. There are three known types of neutrino: electron neutrino, muon neutrino, and tau neutrino, and each can change from one type to another. As you can imagine, their candidacy for dark matter really depends on physicists measuring their masses accurately. Despite the fact that their existence has been confirmed, however, their mass remains unknown. From a theoretical perspective they could have been massless, and even then unless their mass is significant then they will not make a large difference to the total mass of the Universe.

With so much riding on the discovery of the mass of the neutrino, it is not surprising that physicists have made considerable efforts to study them. But as we said earlier, neutrinos are unsociable animals, and rarely interact with other forms of matter. So, how is it possible to study something so evasive? The method is rather simple – at least in principle. It is based on the idea of using a bigger net to catch more fish. Since neutrinos do not often react with other particles, physicists have simply provided more particles with which they can react.

One such experiment – the Super Kamioka Nucleon Decay Experiment, or SuperKamiokande – uses a tank containing 32,000 tons of pure water as a kind of 'neutrino net'. The idea is that as neutrinos fly through the water, the shear volume of water increases the odds of seeing a reaction between a neutrino and atoms of baryonic matter. The principle is that as neutrinos fly through the tank, some will collide with other particles, converting to charged leptons (electrons and muons) in the process. Travelling faster than the speed of light (light slows down in water, but neutrinos do not), the leptons produce an eerie, blue glow known as Cerenkov radiation. To detect this radiation, the tank is lined with 11,200 light detectors, each 50 cm across (the largest in the world), all facing inward at the water. By recording and studying the direction and intensity of the Cerenkov radiation, physicists are able to study neutrino interactions.

Simple as it sounds, the experiment is complicated by the fact that neutrinos are not the only particles that can trigger these flashes of light. Cosmic rays – energetic particles from deep space – can also produce Cerenkov radiation. To minimise this problem, the whole experiment is conducted in one of the most unlikely astronomical sites in the world: 1,000 metres down a mine in Kamiokacho in Japan. But even the kilometre of solid rock surrounding the observatory is not enough protection, and so the tank is enveloped by a second tank filled with 18,000 tons of water which acts as a filter to prevent not only cosmic rays but even the radiation from the surrounding rock from contaminating the experiment. Similar experiments – some of which are still running – have been performed in Italy, Canada, Russia and the United States.

Although the exact mass of the neutrino has not yet been determined, experiments such as SuperKamiokande have determined the lower limit for the mass – and it is pretty small. At this size it is meaningless to express mass in grams or milligrams. As Einstein so famously expressed in his equation $E = mc^2$, mass and energy are equivalent, and so it makes more sense to express the mass of subatomic particles in terms of energy using the electron volt (eV) as the unit of measurement. Particles can have masses of a few eV, many giga electron volts (GeV – 10^9 eV) or many tera electron volts (TeV – 10^{12} eV), and so on. (See Appendix 2 for a full list of abbreviations.) The electron neutrino, for example, is less than 7 eV, and for muon neutrinos and tau neutrinos the upper limits are 0.2 MeV and 18 MeV respectively. While these are quite small numbers – for comparison, an electron is about half a million eV – there were such large numbers of neutrinos in the early Universe that it is still conceivable that they are a significant source of dark matter. If neutrinos have a slight mass, they can form hot dark matter. The sum of the masses of all three types needed to solve the dark matter problem is somewhere in the range 1–30 eV. However, if their mass is any higher than this they would contribute so much to the mass density of the Universe that it would have already recollapsed. The rather obvious fact that the Universe has not yet done this places an upper limit on their masses! The most recent results from WMAP suggest an upper limit of 0.69 eV.

But do neutrinos have any mass at all? There is now extremely strong evidence from the SuperKamiokande experiment that at least two of the neutrinos do have mass, while the third could still be massless. Scientists have extremely strong evidence for the disappearance of the muon neutrino (there should be many more of them than can be detected). The most plausible theoretical explanation for the shortcoming is that the muon neutrino oscillates into another neutrino species. This is a quantum mechanical effect, and the most plausible way that it can occur is if one of the neutrinos has mass. Careful analysis of the data places a lower bound on the mass, which is about 0.008 eV. The experiments conducted so far only reveal the difference in masses rather than the absolute scale of the masses, which is why only a lower bound is possible. It could be higher than that, but if it is 0.1 eV, then neutrinos exist and contribute to the dark matter inventory on the same order of magnitude as is contributed by stars. In other words, neutrinos are just as important to the mass density of the Universe as is luminous matter. However, to play a more significant role in large-scale structure formation – the formation of galaxies and clusters of galaxies, discussed more fully in Chapter 13 – the neutrino mass would have to be quite a lot higher – say between 1 eV and 5eV.

On the other hand, if it were at the other extreme – say 30 eV – there would be difficulties with the way the Universe is structured today. As we shall see shortly, the large-scale structure of the Universe – how the galaxies are arranged in space – is influenced strongly by the amount and type of dark matter. If neutrinos are too energetic, they would have smoothed out the Universe much more than we see today. Even a few years ago, scientists felt that neutrinos with mass in the range of a few eV were required to produce large-scale structure. If a vote were to be taken today, it would probably show that a neutrino mass of even 3 eV, let alone 30 ev, is not popular, for this same reason. Nothing is certain, however, and the role of neutrinos in dark matter is still an intriguing question. It seems almost certain that they have mass – which is a good sign that they play a role in the dark matter story.

Despite this optimism, there are two problems with neutrinos as a source of dark matter. One difficulty is that while they almost certainly have mass, they move very quickly (close to the speed of light). Galaxies cannot therefore easily latch on to large numbers of them; and as we have seen, galaxies are surrounded by halos of dark matter. The other reason is even more bizarre. The Universe is not big enough to accommodate them in the numbers required! Neutrinos belong to a family of particles called fermions, which are the matter particles in the Universe. Fermions include the more familiar particles we know as electrons, protons and neutrons. In contrast to fermions are bosons, the force carriers, which include the photon (particle of light). One of the characteristics of fermions is that they are individuals: no two can be in exactly the same state of motion at the same time. This phenomenon is summarised in the Pauli exclusion principle, and is demonstrated by the existence of white dwarfs and neutron stars. Unlike ordinary stars, which are extremely tenuous spheres of gas, white dwarfs and neutron stars have long since exhausted their supplies of nuclear fuel, and have collapsed in on themselves to form tiny (for a star) and incredibly dense spheres. However, they do not collapse completely, as the particles of matter of which they are made cannot be compressed beyond a certain point. Two fermions cannot be forced into a smaller space once they begin 'rubbing shoulders'. The same applies with neutrinos. Space is big, but it is not infinite, so even when the Universe is filled to the brim with neutrinos there are simply not enough of them to satisfy the dark matter problem. In order to have enough of them there would need to be so many that they would overflow the Universe – which is obviously impossible. Of course, the heavier they are the fewer are required, but current observations indicate that they are neither sufficiently massive nor plentiful.

12

Exploring exotica: WIMPs and axions

WIMPS

If neutrinos are not the answer, what about weakly interacting massive particles – WIMPs? These are a form of cold dark matter, and as the name implies they interact only weakly with ordinary matter – about the same as a neutrino. The masses considered for WIMPs are greater than 10 GeV, with a strict upper bound of about 300 TeV, although most scientists think the upper limit is only a few TeV. There are cosmological arguments that suggest that in a closed Universe there are three zones of mass possible for neutrinos. One of them is somewhere around the level of the ordinary neutrinos discussed above – say 30 eV. Then there is another window where there could be neutrinos of around 3 GeV, and yet a third around 1 TeV. If the masses are not in that region, the result is a Universe that is much too dense – one that is 'superclosed'. These more massive particles are classed as cold dark matter simply because they are so massive and sluggish that they would not come out of the Big Bang with relativistic velocities. They have names such as heavy neutrinos, neutralinos, and so on.

None of these particles has ever been seen, although experiments have been undertaken to try to find heavy neutrinos using detectors made from dense materials such as thalium. Like neutrinos, WIMPs interact only feebly with ordinary matter. They can be thought of as being like slippery cricket balls. They carry quite a wallop, but catching one is difficult, and there is a better chance if it approaches straight on. This is what scientists hope will happen with WIMPs. If a WIMP collides with the nucleus of an atom it produces a kind of nuclear rebound that can be measured. Although it would be electrically neutral, the heavy neutrino would be distinct from the neutrinos known to exist, simply because it is so massive.

FUNDAMENTAL FORCES AND SUPERSYMMETRY

The most popular WIMP candidate is the neutralino – a hypothetical particle about a hundred times heavier than a proton. It is predicted by a speculative idea called supersymmetry – a product of particle physics theory which attempts to bring together the four fundamental forces of nature in the Universe. In doing so it must also unify all the particles of nature. Before looking at supersymmetry and the neutralino it proposes as a solution to the dark matter problem, we will take a brief detour and peek inside the bizarre world of particle physics.

First, the fundamental forces that bind the Universe together. Two of these forces are familiar to us all: gravitation and electromagnetism. Gravitation (gravity to most of us) is often described as the attractive force that all matter exerts on all other matter (although since Einstein it would be more accurate to say it is the distortion of spacetime). Like the other forces, gravitation has a (so far undiscovered) particle called a graviton that carries the force of gravitation between objects. Electromagnetism is a unification of a number of phenomena, including light, radio waves, electricity, magnetism and X-rays. The particle that carries the electromagnetic force is the photon.

The other two forces are less familiar on a day-to-day basis, but are no less important. One is the strong nuclear force that binds protons and neutrons together inside the nucleus of an atom, and also the particles that make up protons and neutrons – the quarks. Without this force, the protons, with a positive charge, would repel each other and blow apart the nuclei of all atoms. This is by far the strongest of the four fundamental forces, and fittingly the particle that carries the strong nuclear force has a sticky name: the gluon. The last of the fundamental forces is the weak nuclear force, which is responsible for such phenomena as (beta) radioactivity. The particles that carry the weak nuclear force have less appealing names: W^+, W^- and Z^0.

These forces vary enormously in strength and range. Gravity, for example, is the weakest by far – some 10^{39} times weaker than the strong nuclear force. Anyone who has ever held a pin with a magnet can see that the gravity of the entire planet is unable to break the electromagnetic bond of a single small magnet. Even the pin itself is testimony to the weakness of gravity, which has even less chance of defeating the forces holding together the atoms in the pin. And yet gravity has by far the greatest reach, able to span the entire Universe and hold it together. In contrast, the strong and weak nuclear forces have extremely short ranges, extending no further than the distances encountered in the nucleus of an atom.

GUTS

Physicists have for a long time sought a way to unite these four forces. In 1968 a major advance was made independently by two American physicists Steven Weinberg and Sheldon Lee Glashow, and the Pakistani Abdus Salam. Between them they showed that the electromagnetic force and the weak nuclear force are interrelated, setting the stage for the yet to be realised unification of the two forces. For this discovery, Weinberg, Glashow and Salam shared the 1979 Nobel

Prize for physics. Since then it has been shown that there seems to be a relationship between the strong nuclear, electromagnetic and weak forces, and this is the goal of the so-called Grand Unified Theory (GUT). There are various versions of this theory, each of which is an attempt to show that the three forces are different aspects of the same uniting force. Importantly, GUTs propose that the symmetry between these forces exists only at very high temperatures – 10^{27} K – attained only at the time of the Big Bang. So far, however, gravitation has yet to join the fold.

SUPERSYMMETRY

Supersymmetry is involved with the difference between energy scales in particle physics. There is a certain fundamental energy scale called the electroweak scale, which sets the strength of the weak interaction and also the masses of all the particles (quarks, leptons, and so on). This mass scale is about 300 GeV. Another fundamental scale is the Planck mass, in which gravitation becomes strong and a theory of quantum gravity is required. This scale is much higher: about 10^{19} GeV. A further generic scale is a hypothetical unification scale of 10^{16} GeV – three orders of magnitude less than the Planck scale, but still many orders of magnitude higher than the electroweak scale. Ultimately, a theory is needed that explains why these fundamental scales in physics differ by so many orders of magnitude. For example, why is the electroweak scale so much smaller than the Planck scale? Supersymmetry is an attempt to explain at least part of this problem.

NEUTRALINOS

Supersymmetry essentially doubles the number of particles in nature. Every particle has a partner called a superpartner: quarks have squarks, electrons have selectrons, photons have photinos, and so on. The neutralino – the superpartner of several particles – is currently a very popular cold dark matter candidate. But that does not mean that it exists; just that it is popular! However, they do have advantages – at least in theory. For one thing, being heavy they can exert an appreciable gravitational influence on their surroundings. Furthermore, theory suggests that they could have been produced in approximately the same numbers as protons and neutrons. All this results in a promising dark matter candidate. (The neutralino is quantal superposition of photino, zino and neutral Higgsino. The zino is the superpartner of the Z boson, and the neutral Higgsino is the superpartner of the yet to be discovered Higgs boson. It is rather complicated, because the neutralino is not the superpartner of any one particle. The authors are not responsible for the terminology.)

WIMP SEARCHES

Neutralinos fit the bill for WIMP candidature, but can they be detected? In order to be dark matter, WIMPs have to have been produced in the Big Bang. They therefore have to couple to ordinary matter with a strength that is roughly the same as the weak interaction, about the same as neutrinos. The most common way of searching for them is by looking for nuclear scattering. WIMPs are expected to interact with protons and neutrons, and thus with nuclei of atoms, and so if they exist they are everywhere. With a sensitive detector that contains nuclei, you might see something mysterious hitting these nuclei, and this might be a WIMP. A collision process is taking place when you suddenly see a nucleus hit for no apparent reason. The technique is to build up enough of these hits to statistically argue that the collisions are due to WIMPs.

Such an experiment is currently being conducted by scientists at the Rutherford Appleton Laboratory, Imperial College and Sheffield University. Deep in a salt mine in Boulby, in the north of England, their detector searches for WIMPs. More than a kilometre below the surface, a thousand workers travel a network of roads that cover 10 km^2, extracting salt and potash twenty-four hours a day. Despite all the activity, the Boulby mine is an ideal place to search for WIMPs. It is the deepest in Europe, and so is able to reduce the cosmic ray interference to a level comparable with the Japanese Kamioka mine. The company that owns the mine – Cleveland Potash Ltd – made available three disused caverns, each the size of a small house. After being dust-proofed, the caverns were fitted out with the experimental and control equipment, linked via lines to a control room on the surface, and from there to the participating institutions. Amid the mining operations, light detectors monitor a 6-kg crystal of sodium iodide for the brief flashes of light expected whenever a nucleus is struck by a WIMP. It is a difficult task, however. For one thing, while the expected rate of neutralino strikes is expected to be 0.01–0.1 per kg per day, background interference is 1,000 to 10,000 times higher. Moreover, neutrons produced by uranium and thorium in the surrounding rock can produce effects that are identical to WIMP interactions.

A similar experiment is being conducted under the Apennines in Italy, with intriguing results. An Italian–Chinese collaboration known as DAMA (DArk MAtter) uses nine 9.7-kg crystals of sodium iodide as targets – again monitoring them for flashes of light. Located in Italy's Gran Sasso National Laboratory, DAMA sits deep underground to maximise its protection against background noise such as cosmic rays. As dark matter particles pass through the detector, some of them will hit nuclei in the detector and produce a signal. As with all of these experiments there is a constant background noise, the origin of which is not well understood. What has drawn attention to them is the observation of slightly stronger signals from the detector in summer than in winter. During the northern hemisphere summer, the Earth is travelling through space in more or less the same direction as the Solar System is moving around the Galaxy. This is the time when you would expect to encounter a greater number of WIMPs, in the same way that if you leaned over the side of a moving boat and moved your hand towards the bow you would experience greater resistance from the water.

This is exactly what the DAMA team claim to have found: around a 10% increase in the number of WIMP interactions during summer than in winter – though admittedly the asymmetry has been detected only at the lower end of the WIMP mass range. Critics have suggested that the possible sources of error invalidate the claim, but the DAMA team remains undeterred. Further complicating the story is that other WIMP detectors have found nothing, despite their being sensitive to the same types of WIMP. Nonetheless, while the researchers admit they have more work to do, they consider the result as encouraging. One way to verify their results is to search for an increase in WIMPs from a source independent of the halo of the Galaxy where the initial signal is claimed to originate. One such source could be a dwarf galaxy in the constellation of Sagittarius. For some time now the Milky Way has been tearing the Sagittarius dwarf apart. The process has produced two arc-shaped tails to the dwarf galaxy, and it just might be that these tails include WIMPs. Because the Earth would encounter WIMPs from the Sagittarius dwarf at different times from those in the Milky Way's halo, the number of WIMP detections should vary at specific times of the year. Furthermore, the energy produced by WIMPs from the halo and the Sagittarius dwarf should also be different. With critics already skeptical of the initial result, it will be interesting to see how this chapter in the story of dark matter detection unfolds.

In contrast with these subterranean experiments, the Cold Dark Matter Search (CDMS) is an experiment involving a large crystal of germanium and operated relatively close to the surface of the Earth. Despite this, it is expected to be even simpler and more sensitive. The CDMS experiment is a collaboration of several institutions, including the Center for Particle Astrophysics – the same that supported the MACHO project. Each time a WIMP strikes a nucleus within the crystal, the nucleus should not only heat up the crystal but also emit enough radiation to ionise hundreds of surrounding atoms. Both the temperature and the ionisation can be detected, indicating the passage of a WIMP. In order to detect the tiny temperature changes, the crystal is kept at a chilly 0.02 K using a cryostat designed and constructed by the Lawrence Berkeley Laboratory. Like similar experiments, the CDMS is conducted underground, but is a mere 10 metres below the surface, in a room excavated below the campus of Stanford University. This provides experimenters with ready access to the equipment, so that as new technology is developed it can be implemented quickly and easily.

Another possibility that may lead to the detection of WIMPs is based on the idea that they are probably attracted to massive objects, such as the core of the Sun or the Earth. Gather enough WIMPs together at one place, and you should start to see annihilations taking place as one WIMP collides with another. A by-product of such annihilations may be energetic neutrinos, and so some experimenters are looking for energetic neutrinos emanating from the Sun. Such neutrinos would be quite distinct from the usual solar neutrinos, because their energies would be much higher. Detectors such as SuperKamiokande are currently monitoring solar neutrinos.

In an attempt to study neutrinos coming up through the Earth, an experiment called AMANDA – the Antarctic Muon and Neutrino Detector Array – is being set up at the south pole. Hundreds of photosensitive devices have been lowered into water-drilled holes that penetrate between 1.5 and 2 kilometres below the surface of the ice. On their way through the Earth, high-energy neutrinos should interact with the surrounding material – ice, rock, and so on – producing a muon. As the muon passes through the ice it will produce Cerenkov radiation, which can be detected and plotted. The reason why this experiment works is that a kilometre below the surface the ice is not white and crystalline like the ice in a domestic freezer. Under the fantastic weight of the ice above it, all the air bubbles have been forced from the ice pack, and the frozen water is as pure and clear and dark as outer space.

WIMP annihilation may also take place in the Galactic halo. We have already seen that the halo of the Milky Way is full of dark matter, and if this dark matter is in the form of WIMPs, then there is again a large concentration of WIMPs ready to annihilate each other. This time the annihilations would produce anomalous cosmic rays, which can be studied when they interact with the atmosphere of the Earth, using strange-looking telescopes such as CANGAROO II – a joint Australian–Japanese facility in the outback of South Australia. Unlike most fields of astronomy, the study of cosmic rays does not require steady skies; but it does need dark skies – and in the South Australian outback the skies are the darkest on Earth (disturbed only by the occasional unwelcome visit by kangaroo shooters and their ever-present searchlights and rifles). Other efforts in this field include the High Energy Stereoscopic System (HESS) observatory in Namibia, and the Whipple Observatory in Southern Arizona.

In contrast with these astronomical pursuits, there is also a purely terrestrial experiment: the detection of a WIMP in a particle accelerator. The Large Hadron Collider now being built at the European Centre for Particle Physics (CERN) will be used to search for supersymmetric particles, including the neutralino. It is just possible that one day one of them might be produced and detected.

AXIONS

The third candidate for non-baryonic dark matter is the axion – yet another example of cold dark matter. Now, while there is confusion over whether neutrinos play a major role in the dark matter story, at least we know they exist; but the same cannot be said of the axion. It is a speculative particle that has been invented to solve a theoretical problem in particle physics, well away from cosmology. If axions exist, they are very light, with a mass limit of 10^{-4}–10^{-6} eV. This is several orders of magnitude lighter than the neutrino, and so if they are the source of dark matter there would have to be an enormous number of them.

For example, if the halo of the Galaxy is full of axions then there would be 10^{13} per cubic centimetre.

There is a component of the theory of the strong nuclear force that binds quarks into protons and neutrons that contains a particular term in the equations that govern it. It is called the θ (theta) term, and it gives rise to the strong 'C-P violation' – a distinction between the interactions of particles and antiparticles. It is cosmologically very important, because it almost certainly plays a role in explaining why the Universe is composed of matter rather than antimatter. The θ term in the strong interaction equations violates C-P in such a way that it really causes a problem. The strength of C-P violation is governed by an arbitrary parameter – a number in the equations which is not set by the theory. Because strong interactions do not distinguish between particles and antiparticles, a very strong upper bound is placed on the value of this arbitrary constant which sets the strength of C-P violation. This number is about 10^{-8}, and so the strong C-P problem is to explain why this number – which could be anything – is so very small. It is one of those fine-tuning problems in particle physics, somewhat akin to the cosmological constant problem but not as severe. It could be said that this is an arbitrary constant which just happens to take this extremely small value. While this is technically consistent, it is perhaps not very satisfactory as an explanation. So, the axion was invented to provide a physical reason why strong C-P violation is very weak or non-existent.

The axion is a boson, which means that it is not a particle of matter but rather a force carrier, like the photon. The masses talked about for axions are very small – about 10^{-5} eV. There are various laboratory and astrophysical constraints on the mass and strength of the coupling of axions to ordinary matter. When the dust settles it is about 10^{-5} eV – perhaps 10^{-6} or 10^{-4}.

Cosmologically, axions were produced during the Quantum Chromodynamic (QCD) phase transition of the Big Bang. In the very early stages of the Big Bang, quarks were liberated; that is, they were free to move around. As the Universe expanded and cooled, a critical point was reached – the QCD phase transition – after which quarks were not free to move around. At this time they began to clump together into baryons – protons, neutrons, and so on. Baryonic matter then congealed out of the plasma which once contained free quarks – which is the reason why it is called a phase transition. It was like water solidifying as ice. If they exist, axions would have been produced at the QCD phase transition, and they should exist today.

Axions can couple with photons, and so one way of searching for them is to watch for their conversion from axions to photons in a magnetic field. The two most important and prominent experiments are carried out at the Lawrence Livermore Laboratory and by the Kyoto group in Japan. The American experiment – begun in 1995 – is a joint effort between the Massachusetts Institute of Technology, Lawrence Livermore National Laboratory, Fermilab, and Lawrence Berkeley National Laboratory. This delicate experiment consists of a 4-metre tall, 12-ton magnet which has been lowered into a hole in the floor of a remote building at the Lawrence Livermore National Laboratory. There, isolated from nearby electronic devices, the $1.4-million device generates a magnetic field about 150,000 times stronger than the magnetic field of the Earth. Within this magnet, a tuneable cavity slowly scans a range of frequencies expected to be equivalent to the mass of axions. If axions exist, when they enter the cavity they should be converted into photons – and that is exactly the object of the experimenters' search. With such tiny masses, the resulting photons would have microwave frequencies: say 0.2–20 GHz. The resulting radiation output would have a very narrow spread of frequencies, about one part in a million. Ultra-low noise microwave amplifiers monitor the cavity so that if and when an axion is induced to convert to microwaves, the event will be detected.

So far, the sensitivity required has not been sufficient, but this negative result has allowed limits to be placed on the mass of the axion and how strongly axions couple to ordinary matter, excluding some, but not all, of the masses that are cosmologically interesting. So while no positive evidence has yet been produced for their existence, they remain a possibility.

This has been a necessarily brief review of non-baryonic matter. We could continue, but it is clear that even though such matter is difficult to detect, its existence is more or less certain. What scientists do not know is exactly what it is. But now we need to return to the question of what role such matter has, or rather had, in the evolution of the Universe.

13

In the beginning...

What role does dark matter play in the Universe as a whole? We have seen that it dominates the dynamics of galaxies and galaxy clusters today, and that whatever it is it outweighs baryonic matter by at least 6:1. But does it have a larger role to play in the future evolution of the Universe, or has it played a major role in the past? The answer is... no and yes. We start, of course, at the beginning: the origin of the Universe.

HOT AND COLD DARK MATTER

In the last chapter we said that dark matter can be divided into two classes: hot dark matter (HDM) and cold dark matter (CDM). The distinction is made because of the different effects each has on the formation of clusters of galaxies way back in the early Universe. As well as explaining the dynamics of clusters of galaxies, dark matter is almost certainly deeply implicated in the formation of large-scale structure in the first place. We have mentioned large-scale structure a few times in this book, but what exactly is it?

On the scale of clusters of galaxies, the arrangement of matter appears to be unique for each region of space. For instance, as we saw in Chapter 3 the Local Group of galaxies consists of two large spiral galaxies – the Milky Way and the Andromeda galaxy – together with a host of smaller elliptical and irregular galaxies. It is unlikely that we will find another cluster of galaxies just like ours anywhere else in the Universe. On larger scales, however – say greater than a few hundred million light-years – the Universe takes on a uniform if foam-like structure. Colossal voids containing relatively few galaxies are surrounded by immense filaments spanning hundreds of millions of light-years. Although these filaments take up only 1–2% of the volume of the Universe, it is where we find most of the matter. And although there are local differences, like bubbles in a bath, from a distance they all look more or less the same. The question becomes: how did this structure form in the first place?

CREATION OF LARGE-SCALE STRUCTURE

This is an important question. The early Universe was quite smooth, yet somehow irregularities were seeded which developed into the structures we see today. The process itself seems simple enough: start with a slightly denser region of space, which because of its slightly greater matter content has a correspondingly higher gravitational pull. This denser region attracts more matter, and so becomes denser and more attractive to passing matter. Bit by bit the denser regions evolve into structures. Meanwhile, the more rarefied regions, where there was less matter, became more and more barren. However, to create the immense structures we see today required enormous amounts of matter; but ordinary baryonic matter could not have accomplished it all on its own, as there is just too little of it to go around. It had to have a helping hand.

According to the leading theory of the early evolution of the Universe, the seeds of the structure we see today emerged during a period of the Big Bang called Inflation. The Inflation theory – first proposed by Alan Guth in 1980 – suggests a period of rapid (exponential) universal expansion which took place around 10^{-35} seconds after the beginning of the Universe. During this brief episode the Universe expanded from being smaller than a proton by a factor of 10^{50} – all in less than the blink of an eye! The whole catastrophic episode was driven by energy released during a period of phase transition. Phase transitions are well known from common experience. Whenever you see steam condensing to water, or water turning to ice, you are witnessing a phase transition – a substance transforming from one state to another. But phase transitions cannot take place without energy being transferred from one place to another. For example, if you place your hand (at a safe distance) above a kettle of boiling water and let the steam condense on your palm, you will feel the release of heat energy into your hand as the water passes from one state (steam) to another (liquid water). A continued transfer of energy out of the water – for example, in your freezer – results in the water turning to ice. In a similar way, the early Universe went through a phase transition, although the consequences and amount of energy released were titanic by comparison.

It was at this stage in the life of the Universe that the seeds of structure were sown. Throughout the phase transition, particles were randomly popping into existence out of the radiation and then disappearing again (see Chapter 8). Known as quantum fluctuations, there were tiny but important irregularities in the density of matter in the Universe. These irregularities were frozen and then amplified by the sudden expansion driven by the phase transition. What is important for us is that the irregularities, being denser than their surroundings, had a slightly greater gravitational pull than the space around them and so attracted more matter. This early clumping of matter grew to the structure we see today.

All this clumping could not happen immediately, however. In the early Universe there was a mixture of 99% dark matter and possibly 1% protons and neutrons (baryonic matter). For a while the baryonic matter and radiation were tied together. Everything was very hot, and radiation constituted a large proportion of the mass of the Universe. (Remember that mass and energy are equivalent.) The episode can be likened to a rain storm. While the storm rages the rain remains largely airborne, and it is difficult to distinguish the water from the wind. In a similar way, because of the intense storm-like conditions of the Big Bang, radiation and matter were more or less the same. In this radiation-dominated epoch, the Universe was expanding too quickly to allow material to fall into over-dense regions quickly enough.

As the Universe continued to expand, the density of radiation and matter were both falling, but the radiation density was falling faster. About 10,000 years after the Big Bang, the density of radiation fell to that of matter, and continued to fall faster. At this stage, cold dark matter began to clump together, but the radiation pressure was still sufficient to keep the baryons flying. Finally, about 280,000 years after the Big Bang the storm began to settle. The Universe expanded and cooled enough so that most of the material became neutral – that is, non-ionised – and radiation and matter went their separate ways. At that point the matter began to 'fall out' of the matter/radiation mix, settling into those places where the gravity was highest – into the tiny irregularities in the density of the Universe that began during the phase transition. The small islands of matter grew, attracting yet more matter, and so on. As the Universe continued to expand, this structure grew with it. The evolution of the structure redistributed the baryons from what was once a rather smooth plasma from only a tiny fraction of the Universe at the time of the Big Bang, to a variety of condensed states.

COSMIC MICROWAVE BACKGROUND

The cosmic microwave background is now more than mere theory. At the time that radiation and matter parted, the seeds of large-scale structure were in place, and these same seeds influenced the paths of the suddenly liberated photons of light. We can still see these photons today as a uniform background of microwaves that blankets the entire sky. This cosmic microwave background (CMB) is predicted in Big Bang theory, and was discovered by Arno Penzias and Robert Wilson in 1965. In 1992, observations made with the Cosmic Background Explorer (COBE) satellite discovered fluctuations in the CMB which revealed themselves as hot and cold spots that differ by a very minute amount (0.00003 K) and ranged in size from a quarter of the entire sky down to an area about the same as your extended hand held at arms length. These fluctuations were evidence of irregularities in the distribution of matter in the early Universe – the seeds of the large scale structure we see today. Since the COBE discovery, dozens of independent high-resolution observations have confirmed and sharpened our view of the cosmic microwave background. But by far the most successful was WMAP, which we mentioned briefly in Chapter 8.

Named after cosmologist David Wilkinson, the Wilkinson Microwave Anisotropy Probe (WMAP) was launched in June 2001 and placed at a stable

Figure 13.1. Where it all began? This image of the cosmic microwave background radiation – the finest and most detailed full-sky map of the oldest light in the Universe yet obtained – was produced using the Wilkinson Microwave Anisotropy Probe (WMAP) spacecraft. The microwave radiation captured in this map is from 380,000 years after the Big Bang, more than 13 billion years ago. The data bring into high resolution the seeds that generated the cosmic structure we see today. These patterns represent tiny temperature differences within an extraordinarily evenly dispersed cosmic microwave background bathing the Universe, which now averages just 2.73 degrees above absolute zero. WMAP resolves slight temperature fluctuations, which vary by only millionths of a degree. (Courtesy NASA.)

point in space 1.5 million kilometres from Earth on the side away from the Sun. With the Sun, Earth and Moon always kept out of its field of view by a protective shield, from its vantage point WMAP produced the most astonishingly accurate map of the CMB. Temperature differences of 0.00002 K were mapped down to a resolution of 0.3 – less than the width of your little finger held at arm's length.

WMAP had one primary purpose: to pin down the free parameters of the Big Bang model of the Universe. Although this model is a good explanation of what we see today, there is, as we have seen, a lot of room for variation. The only way to determine just how the Universe unfolded is by observation of the ancient light from the Big Bang – the CMB. By studying the way the temperature varies with different angular scales, cosmologists can test various fundamental parameters to determine whether they match what WMAP sees. (We shall take a closer look at just what WMAP tells us about the Universe in the final chapter of our story.)

Galaxies and clusters of galaxies are made from baryonic matter, and this matter would have been attracted to the regions of higher density. Since

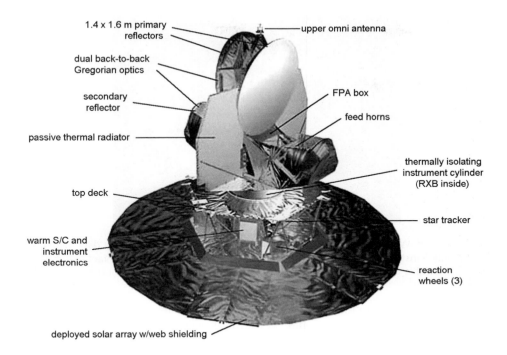

1.4 x 1.6 m primary reflectors

dual back-to-back Gregorian optics

secondary reflector

passive thermal radiator

top deck

warm S/C and instrument electronics

deployed solar array w/web shielding

upper omni antenna

FPA box

feed horns

thermally isolating instrument cylinder (RXB inside)

star tracker

reaction wheels (3)

Figure 13.2. The Wilkinson Microwave Anisotropy Probe. (Courtesy NASA.)

baryonic matter was in such short supply, there would not have been enough of it around to accumulate fast enough to evolve into the structures we see today in the lifetime of the Universe. It had to have a helping hand; and dark matter fits the bill nicely. Remember that dark matter is not influenced by radiation – it is dark, after all – but is influenced by gravity. Dark matter would easily have made its way to any density fluctuations that were around, unhindered by the intense radiation of the early Universe to the denser regions. There it added its weight to the accumulated baryonic matter, helping to attract still more of the material that would one day evolve into the galaxies, stars, and us.

HDM OR CDM?

But which sort of matter was it – HDM or CDM, and would it have made a difference? Observations and simulations of small-scale structure in the Universe hint at HDM, while larger-scale structures favour CDM. To discover what has happened between then and now, cosmologists experiment with the influence of HDM and CDM on the early Universe by using computers to simulate the evolution of virtual Universes. They add various quantities of HDM, CDM, baryonic matter, and so on, press 'play', and sit back to watch the show. The

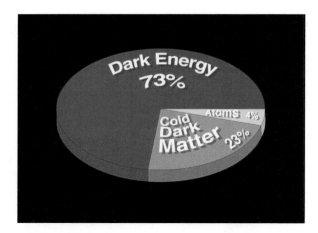

Figure 13.3. What the Universe is made of? After decades of searching for dark matter as the bulk constituent of the Universe, we may now have to settle for the fact that matter of any kind is not even the dominant component of the Universe. (Courtesy NASA.)

verification of a particular mixture lies in whether the virtual Universe evolves into something resembling what we actually see today. As a result of these simulations there are now different predictions from theories of fluctuations in the Universe about what the Universe should look like if it is dominated by hot or cold dark matter.

One of the most important deciding factors is the order in which structure evolved in the Universe. Was it top-down or bottom-up? With top-down, the largest structures formed first, and then the smaller-scale structures fell out of them. Bottom-up is the reverse, in which the smallest structures formed first and accumulated to form large-scale structures.

If the Universe was dominated by hot dark matter, then it is a top-down scenario. The reason is that HDM moves so fast that it tends to spread itself around the Universe rather than being tied down by the tiny islands of baryonic matter. This fits observations of the current Universe very well. While the Universe has structure, on a large scale it is more or less the same everywhere you look; that is, smooth. It is similar to thinking about the surface of the Earth. While it has enormous mountains and deep oceans, compared with the size of the Earth these irregularities are quite tiny. This is now, but back then things were different. Observations made with COBE and WMAP show that small structures formed very early, and not later, in the Universe's history. In the early Universe, HDM particles would be moving far too quickly to ever have accumulated into clumps of the size measured by COBE and WMAP. There is one possible way out of this dilemma, and that is if 'topological defects' such as cosmic strings play an important role in structure formation. However, the party line is that this is improbable.

Cold dark matter, on the other hand, provides a much better medium in which these inhomogeneities can properly develop into fully-fledged large-scale structures. They allow the Universe to evolve bottom-up. CDM particles, by definition, move slowly compared with HDM particles, and at the time structure began to form they were moving slowly enough to be trapped by the gravitational puddles created by baryonic matter. The smaller structures therefore formed first, and slowly accumulated into the larger structures. This scenario is in agreement with observations of the cosmic microwave background radiation.

At the moment, theory does not favour a Universe dominated by hot dark matter; but it should be understood that these theories change almost as rapidly as the transitions that they try to predict, and so it will be some time before anyone can say with absolute certainty which way the judgment of hot or cold dark matter will fall. Nonetheless, the results from WMAP strongly suggest that cold dark matter is the biggest contributor.

So is this the end of the story? Has the dark matter problem finally been solved? Not quite. It seems reasonable that we now know not only where the dark matter is, but also, at last, what it is. Expressed in terms of the critical density, the Universe contains 4% baryonic matter and 23% cold dark matter. And yet scientists still cling to a Universe with a critical density. But if we live in such a Universe, what constitutes the remaining 73%? What is most of the Universe made of? This is the subject of the next and final chapter in our story.

14

Towards omega

Co-author: Charles H. Lineweaver

CRITICAL DENSITY PREFERRED

In Chapter 8 we introduced the idea of the critical density of the Universe – the amount of mass needed to halt the Universe's expansion at some infinitely distant future time provided the Universe is dominated by matter alone. We have seen that baryonic matter constitutes a small fraction of the total gravitational mass of the Universe, and even with the addition of non-baryonic dark matter the Universe still falls short of the critical density. For a long time, many believed it would – and not always for scientific reasons. The critical density is an example of a beautiful symmetry that scientists of all persuasions have come to expect from nature: if flowers and snowflakes, DNA molecules and spiral galaxies form in such elegantly symmetrical ways, why not the entire Universe? It seemed natural to expect the Universe to have a critical density. Furthermore, a popular version of the Big Bang – Inflation – predicts a critical-density Universe for the simple reason that if the early Universe had anything other than the critical density at the time of Inflation – either a tiny amount higher or lower – then the difference would have been amplified tremendously by now. Because Ω_m and Ω_{grav} add up to 0.27, and this is so close to 1 (considering the alternatives), many believe that the Universe has indeed the critical density. Surely it was simply a matter of looking hard enough to find the rest of the Universe; but as we have seen, astronomers have been looking for a long time and have found nothing. Will the Universe one day halt its expansion due to self-gravitation? This question has been answered. It contains nowhere near enough matter to do anything of the sort. Matter may be important to us, but it was a disappointment for those expecting it to be the dominant force in the Universe. The Universe will continue expanding forever.

AN ACCELERATING UNIVERSE: DARK ENERGY

No sooner had everyone accepted that the Universe would continue to expand than evidence emerged that the rate of expansion was increasing. The news came in the form of observations of supernovae – the explosions that mark the end of life of massive stars (as discussed in Chapter 3). Supernovae have long been used as standard candles. A number of these standard candles are visible at various distances throughout the Universe – stars that vary in brightness according to predictable patterns, galaxies of known type and size, and so on – but each is too faint to be seen or identified across the ocean of space. When supernovae explode they shine as brightly as a hundred billion stars – as bright as an entire galaxy. Such a spectacle does not last long, however. Within days they begin to fade, and after a few weeks all but the nearest will have faded from view, but while they shine they are beacons that indicate, among other things, how far away they are. Supernovae come in different flavours – each of them distinguished by how it brightens and fades over time, and each with its own true maximum brightness. By monitoring the behaviour of a supernova in a distant galaxy, astronomers can determine what type it is and how bright it would appear if it were nearby; and by comparing the apparent and true brightness of the supernova, can determine its distance and thus the distance of the host galaxy. This information can be combined with the redshift. As discussed in Chapter 7, the amount of redshift in the spectrum of an object at a cosmological distance indicates how fast the object – and its region of space – is receding.

By comparing the recession velocities of objects at various distances we can determine how fast the Universe is expanding and whether the expansion is changing with time. Even though there may not be enough gravity to stop the Universe expanding forever, the very fact that matter exists means that gravity should at least be putting up a fight against the expansion. Expand forever it might, but the Universe should be slowing down a little.

In 1997, results of an observing run at Cerro Tololo Inter-American Observatory in March of that year began to filter through to the astronomical community. A team of astronomers lead by Saul Perlmutter, of the Lawrence Berkeley Laboratory in California, had studied the spectrum of supernovae the most distant of which is some 5 billion light-years from the Earth, halfway back to the Big Bang. Their announcement was followed shortly afterwards by the results of work by Brian Schmidt and his colleagues at Mount Stromlo and Siding Spring Observatories, who also arrived at the same conclusion by studying supernovae. Rather than slowing down, the rate of expansion of the Universe is increasing, indicating that the expansion is being driven by some unknown form of 'dark energy'.

This dark energy has been given a name: Ω_Λ – the 'cosmological constant'. Einstein himself originally incorporated such a cosmological constant in his original version of General Relativity. Although his theory predicted either an expanding or contracting Universe, at the time (1929) the entire 'universe' consisted of the Milky Way, and there was no evidence that the Milky Way was either expanding or contracting. He therefore added the term to force the theory to predict a static Universe. As is well known, when Einstein finally learned of Hubble's discovery of the expansion of the Universe, he retracted the cosmological constant, referring to it as the greatest blunder of his life. Now the cosmological constant has returned – and returned in force, with excellent observational evidence to support it.

WHAT IS THE COSMOLOGICAL CONSTANT?

But what could the cosmological constant be? What could make space expand faster – not only resisting the inward pull of gravity, but overpowering it? A satisfactory theoretical explanation of the cosmological constant has not yet been proposed, but the leading model is that it is the energy of the vacuum of space. One of the reasons why this theory is so interesting is that it is one of the few places in physics where General Relativity and particle physics come together. It is a parameter inside General Relativity, which is a classical theory (that is, it does not have quantum mechanics in it), and yet its origin is supposedly quantum mechanical: the energy density of the vacuum. At least, that seems to be the standard interpretation of it; but from what particle physicists know of the vacuum energy, they have so far been unable to explain the mechanism behind it.

WHAT IS VACUUM ENERGY?

Exactly what is 'vacuum energy'? How can 'nothing' drive apart everything? If all the air is extracted from a bell jar then it is like the vacuum in space, though in space it is an even more rarefied vacuum. In any case, it should be treated as if nothing is there. But in that nothing – in that vacuum – there is a seething froth of virtual particles that can be measured indirectly via a phenomenon called the Casimir effect.

The Casimir effect is named after the Dutch physicist Hendrik Casimir, who proposed the following experiment. Place two metal plates very close together in a vacuum. Pairs of virtual particles will continue coming in and out of existence on either side of the plates as well as between them. Virtual particles come in a whole range of so-called Schrodinger wavelengths – some long, some short. As the plates are brought closer and closer together, the wavelengths that are longer than the distance between the two plates will eventually be cut off, resulting in fewer virtual particle pairs coming into existence between the plates than outside them. The pressure of the vacuum between the plates is consequently less than that on the outside of the plates, and so the plates are pushed together. Tiny as it is, this force – the vacuum energy – can be measured. It is a quantum effect – it has nothing to do with gravity or electrostatic forces or anything else – and so is

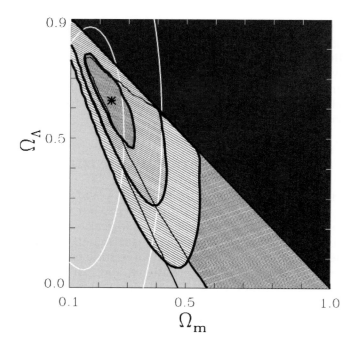

Figure 14.1. The first observationally-based determination of Ω_Λ (0.620.16) and Ω_{matter} (0.240.10) with error bars. The current best values of Ω_Λ and Ω_{matter} are within these error bars. The thick dark lines are the approximate 68.3%, 95.4%, and 99.7% contours from the joint probability of the CMB and non-CMB constraints. The 68.3% and 95.4% confidence levels from non-CMB observations are in white, and the 68.3% and 95.4% confidence levels from CMB observations are the thin black lines. (*Astrophysical Journal*, **505**, L69–L73 (1998).)

an indirect measurement of these virtual particle pairs. This experiment has been carried out repeatedly, and has verified the existence of vacuum energy.

Even if the vacuum of space has a negative pressure (antigravity) that is blowing up the Universe, how can it continue even when the Universe is such an immense size? The answer lies in the fact that vacuum energy is not diluted by the expansion of the Universe. Quite the reverse. Normal matter is diluted with the expansion of the Universe: galaxies, clusters of galaxies, and the vast quantities of dark matter all have to exert their gravitational influence over ever increasing volumes, robbing them of their power. But the structure of the vacuum is not weakened by the expansion of space. No matter how big space becomes, the density of the vacuum energy stays the same. Moreover, with more space there is more vacuum, and so the total vacuum energy increases.

COSMOLOGICAL PARAMETERS

We are now faced with the problem of firmly establishing three of cosmology's parameters: Ω_b, normal baryonic matter; Ω_{cdm}, cold dark matter; and now the newcomer, Ω_Λ, the cosmological constant. As we have seen, the accepted values for these three parameters have changed as newer and more accurate observations have become available. A few years ago a survey might have produced the following popularity figures: 10% favoured the Einstein–de Sitter model in which the amount of mass is equal to the critical density and there is no cosmological constant – $\Omega_m = 1$; 30% favoured an open Universe in which Ω_m is 0.2–0.3 and $\Omega_\Lambda = 0$; and perhaps up to 60% favoured an Ω_m of 0.2–0.3 and an Ω_Λ of 0.7. The first model, in which Ω_m equals the critical density, is favoured by only a minority and is almost dead. This leaves two possibilities: one in which the Universe contains matter and not much else (this model is not any better than the first one), and the more recent contender in which matter makes up 27% of the critical density and the cosmological constant makes up the remaining 73%.

CONSTRAINING Ω_b, Ω_m AND Ω_Λ

Evidence for the cosmological constant has come from a variety of sources, but the strongest support comes from comparing various constraints from those sources. Independent constraints are important here. No matter how well an observation is made nor how esoteric the theoretical consideration of the results, there is always a range of possibilities that can explain them, each with their own probability of being correct. For example, if you were to see a meteor streak through the night sky you could plot its course from where you were standing and make a rough prediction of where it fell. But factors such as apparent brightness, speed, unknown altitude, and so on, all influence the range of possibilities of where the meteorite actually fell. At best, all you would be able to say is that it fell somewhere within a given area; that is, within certain geographical limits. These limits are constraints, and while the closer to the centre of the predicted fall area, the more confident you are of finding your 'fallen star', there is still no guarantee of finding it there. One way out of this dilemma is to compare your prediction with that of a friend who also saw the meteorite fall, but from their backyard several kilometres away. Now you have two sets of constraints to compare, and where they overlap there is a much higher probability of finding the meteorite. Add a third set and your confidence increases even more, and so on. But there comes a point at which you have to consider that science cannot ever prove anything, but we are confident enough to move on from here and go looking for the meteorite without wasting our time.

In the 1990s this same approach was taken by Charley Lineweaver, then at the University of New South Wales, in his search for the value of the cosmological constant. Lineweaver was one of the first to compare the constraints from observations from a variety of sources, including studies of the cosmic microwave background radiation, distant supernovae, and mass–luminosity ratios of clusters of galaxies. From these comparisons he constrained the value of Ω_Λ where all the observations tend to agree in parameter space, and derived the figures $\Omega_m = 0.24$ and $\Omega_\Lambda = 0.62$.

Most recently, more precise results have come from WMAP. Between 1998, when Lineweaver's results were published, and 2002, when the WMAP results appeared, the constraints on Ω_Λ and Ω_m and the Hubble constant had progressively improved. But by producing such a detailed high-resolution, high-sensitivity map of the CMB, WMAP has provided cosmologists with a way of refining these figures still further (in some cases reducing the error bars by a factor of two or so), and it was able to do this simultaneously with about six parameters. Since the results of this and other surveys of the CMB are so important to our understanding of the Universe, it is worth spending a short time looking at how the data are interpreted.

Remember that the inhomogeneities in the distribution of cold dark matter were laid down after the density of matter exceeded that of radiation around 10,000 years after the Big Bang. Shortly after that (380,000 years), radiation and matter went their separate ways – matter and radiation 'decoupled' – and the radiation streamed through the Universe to arrive at us today in the form of the CMB. The relative intensity of the CMB has much to tell us about the structure of the Universe at the time of decoupling. The situation has been likened to sunlight being scattered through clouds: what we see of the light comes to us from the nearest surface of the clouds, the surface of last scattering. Regions with larger matter content would have had greater gravity, and hence a greater pull on the photons leaving them. This makes them appear fainter, or cooler, than average. A study of the variation of the CMB at the larger scales therefore provides an indication of the distribution of dark matter at the time of decoupling.

At smaller scales, variations are caused by sound waves that were probably induced by gravitational fluctuations created during the Big Bang. Travelling through the early Universe they caused regions of denser and rarer matter; and where the matter was denser there were more photons. The fluctuations in the CMB therefore reveal how far they travelled before decoupling. Hotter, brighter regions indicate denser matter, and so on. Since the fluctuations we see came from a time 380,000 years after the Big Bang, that is how long the sound waves had to travel through the Universe before their effect on the gas was frozen into place. Thus the distance that they travelled – the 'sonic horizon' – provides a fundamental distance scale for the early Universe. Furthermore, since they were travelling through a fluid of hydrogen, a study of them provides an indication of the relative density of protons and electrons (the two particles that make up hydrogen). Varying the density changes the speed with which sound travelled through the early Universe, which is also indicated in the fluctuations of the CMB.

By interpreting the data, cosmologists have identified the contents of the Universe. The results are: $\Omega_{baryons} = 0.04$, $\Omega_m = 0.27$, and $\Omega_\Lambda = 0.73$. Combining these figures produces $\Omega_{total} = 1.0$ – the long-suspected critical density.

COULD THIS BE IT?

Is that it? Has the riddle of dark matter finally been solved? It would be foolish to answer 'yes' to this question, or to a similar question about any other line of research about our Universe, and it is certainly too early to say that we have identified the nature of dark energy. This is not to say that astronomers' methods are wrong or that their reasoning is flawed; simply that the Universe is a huge, complex and ancient place, and we have been studying it for an extremely short period of time. Even so, by the time this book is published it is probable that further refined figures from WMAP will have emerged, but these are not expected to differ much from those quoted here. We have indeed entered the age of precision cosmology.

It is reasonable to assume that we have as much to learn about the Universe as we ever had. While some feel we are close to discovering the last of the big pieces of the puzzle, the truth is that astronomy and cosmology are part of a wonderful journey – an exploration of an increasingly complex Universe of which we are a part. From time to time it may seem that we are only a few steps away from making sense of it all. The confirmation of the amount and nature of dark matter in the Universe is an example of one such giant stride. But the 80-year-old riddle of dark matter has now been joined by an even bigger puzzle: the nature of dark energy. The experience serves only to remind us that no matter how sure we feel about our understanding of the nature of the Universe, from time to time something unexpected will always come out of the darkness to surprise us.

Appendix 1

What is matter?

What are we humans, animals, plants, all forms of life, the Earth, the planets and the stars really made of? And is that familiar stuff really all there is to the Universe? These are deceptively simple questions that go right to the heart of the quest for dark matter. In this book we have explored different aspects of matter and its partner, energy, and have considered why some types of undiscovered matter are suspected so strongly that scientists have invested huge amounts of money in machines to look for them. This is the story of the origin of the Universe and how it led to the types and amounts of matter we see today. Here we review the basic ideas on the nature of matter, what it is made of, and how it is all put together.

DEFINITION OF MATTER

The first thing to understand about matter is that not all of it can be seen directly, and not everything you see around you is matter. Common examples of this are air and light. We can feel air, and even see its effects on things such as clothes on a washing line; but generally speaking we cannot see it – at least not without specialised equipment. In contrast, light allows us to see everything around us. Every sunrise we are bathed in it – and life would soon come to an end without it. But light is not matter; it is energy.

So, what is matter? There are several criteria that must be satisfied before an entity can be described as matter. First, it must occupy space. You are made of matter, and you can prove that you take up space each time you climb into a bath full of water. The rising level of the water is evidence that you have volume. Less obvious is that you resist change in motion. Perhaps a better example is a broken-down car. Faced with the need to push it off the road, you encounter inertia – the resistance to change in motion, even if that motion remains still. Admittedly, the friction of the tyres on the road provides further resistance, but even if you were to try pushing a car on an ice rink you would find it less likely to move than you, because it has more inertia than you do. In other words, there is more car than you.

Finally, matter has mass. Mass is defined as the amount of matter in an object, but it should not be confused with weight. The familiar example of comparing weight on the Moon with your weight on the Earth may temporarily make you feel better – but your mass and volume remain the same in either environment.

Matter is therefore anything that can be weighed, pushed, pulled, have its shape changed, and so on. This is intuitively obvious, but needs to be directly stated. What is far less obvious is the nature of matter. If we were to delve down into its very fabric, what would we see? Around 500 BC the Greek philosopher Democritus gave considerable thought to this, and in his mind saw tiny, indivisible particles that make up all matter. He called them atoms – meaning 'not able to be cut'. As smart as Democritus was – he was, after all, correct – he was only speculating. He had no evidence to support his ideas, and at the time, many other people disagreed with him. These days, however, we can not only delve into the world of matter to see what is there, but we have evidence to support our conclusions about this bizarre world. Some of the evidence is as beautiful as the revelations in this book, while some of it – such as nuclear bombs – is terrifying. The point is that we are now closer than ever before to understanding the true nature of matter – at least the matter we can see, and perhaps even matter that we cannot see.

MACROSCOPIC: CELLS

Consider a convenient example of matter: the pages of this book. Look closer, past the words, the letters and the full stop at the end of this sentence. Look at the fibres that make up the pages – fibres that were once living tissue in trees. That tissue is made of cells, and those cells are made of countless indescribably complex parts – membranes, cytoplasm, mitochondria, and so on – that once carried out the life processes now ended. Those parts are made up of some of the largest and most complex molecules in the Universe – a familiar example being the molecule of life, DNA, that carries the genetic code enabling us and all living things to reproduce.

MOLECULES

Molecules come in an astonishing variety of forms, and most of the material things that impact on our lives are composed of different types of molecule. For example, water, without which life would cease. It is a molecule consisting of two types of even smaller particles called atoms. Atoms come in different flavours, called elements, of which about a hundred are known. The two elements in water, for example, are oxygen and hydrogen. Elements combine to form all the molecules around us, and react together to form new substances in chemical reactions. Common examples are encountered when cooking, for instance, or when petrol combines with oxygen in your car. In both cases, there is an exchange of energy, but not of matter. In the case of cooking, energy is input so that the various molecules react and make the cake rise.

Inside a car's engine, on the other hand, an initially small input of energy (the spark from the spark-plug) triggers a reaction between the petrol and oxygen to release energy, which drives the car. In both instances, new combinations of elements and compounds are produced, but matter is neither created nor destroyed.

ELEMENTS

The entire Universe consists almost exclusively of two elements: hydrogen and helium. All the things that you and I find so interesting on a daily basis (with the exception of dark matter, of course) are made up of elements that are in the extreme minority. But all these elements – from the lightest and simplest (hydrogen) through to elements with heavy masses and heavier names such as ununnilium – are all made of atoms. But these are not the atoms conceptualised by Democritus.

ATOMS

Atoms are in turn made of four things: protons and neutrons making up a nucleus, electrons surrounding the nucleus, and a large amount of space. But how much space? Imagine that the nucleus of the atom is the size of a walnut. At this scale the electrons would be whizzing around in a sphere more than a kilometre across. Despite this immense space, 99% of the matter in an atom is in the tiny nucleus. So what is holding it all together? The answer is energy: the electrical charge of the electrons is negative, and that of the nucleus (for reasons we shall see in a moment) are positive. The negative and positive charges attract and hold the electrons in place. But when it comes to atoms encountering other atoms, they repel each other for the simple reason that they have electrons on the outside – and two negative electrons repel each other. This property allows you to hold this book. The negatively charged electrons in the atoms and molecules of your hand are repelling the electrons in the atoms and molecules of the book. If this mutual repulsion did not take place, not only would the book literally slip through your fingers, but you would slip through the chair and then into the Earth. This is not as silly as it sounds, for there are particles of matter in the Universe that do just that. We will now continue our downward plunge into the essence of matter.

STRUCTURE OF THE ATOM: ELECTRONS, PROTONS AND NEUTRONS

In the very nucleus we find two more types of particle: protons and neutrons – the heavy components with all the mass. Protons have a positive charge – the same charge that attracts the electrons and stops the atom falling apart. Neutrons have no charge at all, but nonetheless are held in place inside the nucleus by another force different from the electrical force: the strong nuclear force.

QUARKS

Even more fundamental than protons and neutrons are the particles of which they are made: quarks. There are different types of quark – up, down, strange, charmed, top and bottom – that combine in different ways to make up matter. Protons, for instance, are made of two up quarks and a down quark, while the neutron is made up of two down quarks and an up quark. The names of the quarks have absolutely nothing to do with their properties, and are merely labels.

LEPTONS

What about the electrons? What are they made of? Electrons are part of a family of much lighter particles called leptons. Other members of the lepton family include the muon and tau particles, and each of these three has an associated particle called a neutrino: electron neutrino, tau neutrino, and muon neutrino. Neutrinos are strange beasts which can pass through ordinary matter as if it did not exist. This presents some interesting challenges for physicists intent on not only detecting them but trying to measure their mass. Leptons do not appear to be made of anything other than leptons; that is, they show no signs of having any internal structure.

There are, moreover, many other particles, and there are more than two hundred either known or predicted.

ENERGY

Ask a physicist what energy is, and the answer will be that it allows work to be done. It is an elusive concept, and although we can describe its properties, classify it in different ways, and predict not only its behaviour now but its relationship with matter at the beginning of the Universe, the nature of energy remains a mystery.

Energy is the capacity to do work – to change things. The following are some of the more commonly encountered types, although there are many variations:

- *Kinetic energy* Movement. Ride a bike, dance a waltz, fly to the Moon, and you are experiencing kinetic energy.
- *Potential energy* Stored energy, which manifests itself in a variety of forms such as chemical and gravitational energy. Food, petrol and water at the top of a cliff, about to fall, all have potential energy. One of the wonderful things we have learned to do over the millennia is to store energy in

everything from food silos that feed us to mini nuclear reactors that power spacecraft on their interplanetary voyages.

- *Thermal energy* Heat – the amount of energy contained within the excitedly moving particles that constitute matter. The faster the particles are moving, the more thermal energy they possess, and in sufficient numbers they can produce a tremendous amounts of energy. Although thermal energy is referred to as heat, an object with a large amount of thermal energy is not necessarily hot, while a hot object does not necessarily contain a lot of thermal energy. A swimming pool of warm water, for example, contains vastly more thermal energy than a red-hot nail, simply because there is a lot more water than nail!

- *Electrical energy* One of the most familiar forms of energy – for the very good reason that we use a large amount of it! It is a flow of electrons, which are amazingly simple to transport once the infrastructure is in place. Once electrical energy is delivered to our homes we can convert it into many other forms of energy.

- *Radiant energy* The most familiar example is light, but there are other forms: gamma rays, infrared (used for heating and in domestic remote-control devices), microwaves (for communications and cooking), X-rays (medicine), radio waves, and ultraviolet. All of these are part of the electromagnetic spectrum.

Each of these forms of energy (and those not mentioned here) fall into one of four fundamental forces in nature, which are carried from one place to another due to massless particles called bosons. These forces and their carriers are:

- Electromagnetic force, carried by the photon.
- Gravity, carried by the graviton.
- Strong nuclear force, carried by gluons.
- Weak nuclear force, carried by the W^+, W^- and Z^0 particles.

The importance of the relationship between matter and energy is that it is interchangeable: Matter is converted to energy inside stars, for example, and it is this conversion that allows us to continue living. But for the purposes of our story, it is important to understand that the Universe contains a certain amount of matter and a certain amount of energy, and, as we have discovered, a certain amount of dark matter. And that is it. No more, no less. In the history of the Universe there have been some bizarre conversions between matter and energy, and no doubt dark matter, but the total remains the same. One of the goals of cosmology is to determine how much of each – matter, energy and dark matter – exists in the Universe, simply because this will help reveal fundamental properties of the origin, evolution and fate of the Universe in which we live.

Appendix 2

Expressing mass

The mass of subatomic particles is expressed in electron volts (eV). The following are standard prefixes:

Multiple	Prefix	Symbol
10^3	kilo	keV
10^6	mega	MeV
10^9	giga	GeV
10^{12}	tera	TeV

Index

Printing: Mercedes-Druck, Berlin
Binding: Stein+Lehmann, Berlin